Władysław Homenda, Witold Pedrycz

Automata Theory and Formal Languages

Also of Interest

Analog Computing
Development, Programming, Applications, and Future
Bernd Ulmann, 2026
ISBN 978-3-11-914628-9, e-ISBN (PDF) 978-3-11-221017-8

Mathematics of Deep Learning
An Introduction to Foundational Mathematics of Neural Nets
Leonid Berlyand, Pierre-Emmanuel Jabin, 2026
ISBN 978-3-11-914411-7, e-ISBN (PDF) 978-3-11-221821-1

Mobile Applications Development with Python in Kivy Framework
Tarkeshwar Barua, Ruchi Doshi, Kamal Kant Hiran, Avinash Sharma, 2025
ISBN 978-3-11-129319-6, e-ISBN (PDF) 978-3-11-129353-0

Languages and Automata
GAGTA BOOK 3
Edited by Benjamin Steinberg, 2024
ISBN 978-3-11-099642-5, e-ISBN (PDF) 978-3-11-098432-3

De Gruyter Series on the Applications of Mathematics in Engineering and Information Sciences
Edited by Mangey Ram
ISSN 2626-5427, e-ISSN 2626-5435

Władysław Homenda, Witold Pedrycz

Automata Theory and Formal Languages

From Regular Grammars to Turing Machines

2nd Revised and Expanded Edition

DE GRUYTER

Authors

Prof. Władysław Homenda
Warsaw University of Technology
Faculty of Mathematics and Information Science
ul. Koszykowa 75
00-662 Warsaw
Poland

The John Paul II Catholic University of Lublin
The Faculty of Social and Technical Sciences
Al. Racławickie 14
20-950 Lublin
Poland
wladyslaw.homenda@pw.edu.pl

Prof. Witold Pedrycz
University of Alberta
Dept. of Electrical & Computer Engineering
ECERF Building
Edmonton T6R 2V4
Canada

Polish Academy of Sciences
Systems Research Institute
ul. Newelska 6
01-447 Warsaw
Poland
wpedrycz@ualberta.ca

ISBN 978-3-11-914199-4
ISBN 978-3-11-222312-3 (PDF)
ISBN 978-3-11-222377-2 (E-PUB)
DOI https://doi.org/10.1515/9783112223123

Library of Congress Control Number: 2026934496

Bibliographic information published by the Deutsche Nationalbibliothek
The Deutsche Nationalbibliothek lists this publication in the Deutsche Nationalbibliografie; detailed bibliographic data are available on the Internet at https://dnb.dnb.de.

De Gruyter and Walter de Gruyter GmbH are part of De Gruyter Brill.
www.degruyterbrill.com

Questions about General Product Safety Regulation: productsafety@degruyterbrill.com

Cover illustration: Suebsiri / iStock / Getty Images Plus

Contents

Part II: Automata and accepting languages

Foreword

What is Automata Theory and what are its relations to formal languages? What can they do? How do they work? How can one use them? Those are the questions I posed in my Foreword for the first edition of the treatise by Władysław Homenda and Witold Pedrycz. This timely book authored by them provides the reader a systematic exposure to the answers to those questions. As the first edition, the book is very easily understood, without oversimplification of the subject matter. The exposure of the material is concise, yet precise. A clear, step-by-step Introduction to Formal Languages and Automata Theory addresses the most important elements in the field.

To view the developments in this field, a bit of personal history is worth highlighting. When Sam Lee and I wrote our book 'Fuzzy Switching and Automata: Theory and Applications' in 1979, our expectation was the applicability of the contents. However, it is the seminal work of many researchers since then that showed how Automata Theory and Formal Languages could be translated into a working reality. It is within the past few years that it has become increasingly clear that a symbiotic relationship between this field and AI, for example, as well with a variety of techniques from Uncertainty Management can be exploited to conceive, design, and build systems and products having high MIQ (Machine Intelligence Quotient). With the advent of LLMs, there are new opportunities and uncharted territories of applications of automata theory.

Viewed with those recent developments, the importance of the present manuscript is easily understood. Professors Homenda and Pedrycz have assembled an impressive authoritative treatment of some of the basic issues arising in the field. This edition of the book brings a new material covered in Chapters 10–12. Chapter 10 delivers expand on the selected topics of the Turing machines. Among them two-dimensional machines are proved to be equivalent with one-dimensional ones. It is an interesting intuitive confirmation of the Church hypothesis: since Turing machines are the most general and powerful computational models, then their modifications must be equivalent. Chapter 11 is focused on context-free structures. It covers a practical method of analysing expressions (relational, logical, arithmetical). It is also a key method of analysis, compiling and processing them in context of programming languages. Chapter 12 brings forward the essential facets of finite automata with fuzzy sets and elaborates on ways of coping with imperfect information and processing time series.

Professors Homenda and Pedrycz deserve much credit as well as our congratulations on producing an excellent text with authority, originality, and skill, which make this work a pleasure to study and read, and it is only my hope that this outstanding text will serve as an impetus for continued interest in the study and research in this exciting field.

Tampa, Florida, January, 2026

Abraham Kandel
Distinguished Professor Emeritus
Dept. of Comp. Sci. and Eng.
The University of South Florida

https://doi.org/10.1515/9783112223123-202

Preface

Automata theory and formal languages are the cornerstone of computer science and all information processing pursuits. These fundamentals are at the center of any computer science, computer engineering and information technologies curriculum. Writing a textbook in this area is a challenging endeavour. The area is well established yet the ongoing technological trends and the advancements in application areas call for a creative exposure of the fundamental material. This textbook is aimed at students in senior years of undergraduate programs and the first year of graduate programs (both at the MSc and PhD level).

When writing this text, we strived for a systematic and clear exposure of the key ideas by positioning them in the context of the overall picture and a spectrum of possible applications.

There are several outstanding features that will appeal to a broad spectrum of instructors and students, which make this text unique in comparison with the books available on the market:

(i) material is prudently selected to cater to a spectrum adequate for course well adopted by the audience,
(ii) material is made self-contained and motivated,
(iii) the length of the text makes it ideal for a one-semester course by focusing on the required well-structured and motivated ideas,
(iv) in-depth explanation supported by a wealth of solved examples and problems,
(v) a systematic exposure of the material,
(vi) easy expansion facilitating the usage of the text in a variety of courses.

The textbook could be also regarded as a sound and systematic prerequisite to more advanced subjects in the areas of advanced algorithms, computability and computational complexity.

The book is structured into four parts preceded with all required prerequisites offering a concise and focused background to make the exposure of the material self-contained. The four main parts are structured to reflect to the general Chomsky hierarchy of languages, namely:

(i) we start from the class of regular languages forming the simplest class of simplest languages, and then move towards the class of recursively enumerable languages, the most complex class of languages (Chapters 2, 3 and 4). We cover tools that generate languages: regular expressions and regular grammars (Chapter 2), context-free grammars (Chapter 3) and context-sensitive and unrestricted grammars (Chapter 4),
(ii) we go back to the Chomsky hierarchy by studying tools used to accept languages starting from the most complex class of languages and ending with the simplest class of languages (Chapters 5, 6, and 7). Here considered are Turing machines with their diverse categories, Turing machines with stop property and linear bounded automata, which accept recursively enumerable languages, recursive languages

https://doi.org/10.1515/9783112223123-203

and context-sensitive languages (Chapter 5). Then we discuss push-down automata, which accept context-free languages (Chapter 6). Finally, we study finite automata, which accept regular languages (Chapter 7),

(iii) then, we revisit both paths by unifying tools to generate and accept languages (Chapters 8 and 9),

(iv) finally, we go along the Chomsky hierarchy presenting selected advanced topics concerning Turing machines (Chapter 10) as well as some real-world applications of context-free grammars (Chapter 11) and finite automata (Chapter 12).

The text is structured in a way it delivers a great deal of flexibility and makes the material suitable for a broad spectrum of possible courses depending upon the audience and the assumed position of the course in the curriculum. The roadmap shown in Figure 1 displays a number of possibilities highlighting how the material could be covered:

(i) Chapters 2–7 are independent from each other; the reader can study selected ones in any order. In general, minor linkages between Chapters 2–4 and between Chapters 5–7 are not detrimental to the individual studies,

(ii) the remaining chapters require the mastery of the knowledge of previous Chapters.

It is worth to draw attention to Chapter 5, where the concept of nondeterminism is introduced and extensively discussed. This concept is the only obstacle prerequisite to study some topics covered in Chapters 6 and 7.

While the exposed of the material is self-contained, we assume some familiarity of basic concepts and methods of algebra, set theory and logic, and algorithms and data structures. These concepts are included in Chapter 1. It is recommended to revise these concepts prior to studying the remaining parts.

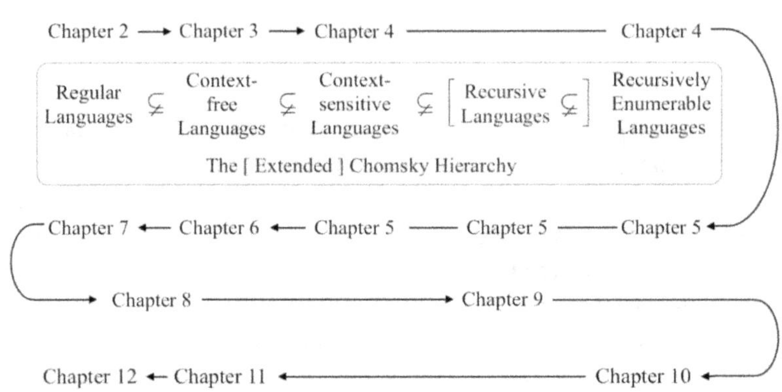

Figure 1: Organization of the material.

We hope that the textbook will appeal to a broad audience of undergraduate and graduate students and all of those interested in the systematic and concise exposure to the fundamentals of automata and formal languages.

Warsaw, Edmonton, January 2026

Władysław Homenda
Witold Pedrycz

1 Preliminaries

This chapter offers all required prerequisites and elaborates on the notation used throughout the book. As such, the content presented here becomes helpful in the systematic exposure of the material covered in the consecutive chapters.

1.1 Sets

Finite sets include a finite number of members. The empty set, that is, the set which includes no members, is also finite. The empty set is denoted by $\emptyset$. Infinite sets include infinitely many members. Infinite sets are countable or uncountable.

We say that a set is countable if and only if its members could be arranged in a particular sequence. In other words, a set is countable if and only if we can enumerate (list) its members assigning natural numbers to them. Such an arrangement guarantees to find a number assigned to any of its members. Of course, finite sets are also countable. Thus the conclusion that any subset of a countable set is countable becomes obvious.

A set is uncountable if and only if there is no enumeration of its members. Any set including an uncountable subset is uncountable.

Example 1.1. Here are some important sets of numbers:
- $P = \{2, 3, 5, 7, 11, 13, 17, \ldots\}$ – the set of prime numbers,
- $N = \{0, 1, 2, 3, 4, 5, 6, 7, 8, \ldots\}$ – the set of natural numbers,
- $Z = \{\ldots, -3, -2, -1, 0, 1, 2, 3, \ldots\}$ – the set of integer numbers,
- $Q = \{a/b : a, b \in Z, b \neq 0\}$ – the set of rational numbers, which are fractions of integers with nonzero denominator,
- R – the set of real numbers. It includes rational numbers as well as irrational numbers as, for instance, π, e, $\sqrt{2}$.

Note that all but the last set of numbers are countable. The last one is uncountable. The following notes concerning the above sets show arrangements of their elements in sequences, that is, justify countability of these sets:
- comparing Example 1.1, it is straightforward that the set of prime numbers and the set of natural numbers are countable,
- the following enumeration of the set of integer numbers $Z = \{0, -1, 1, -2, 2, -3, 3, \ldots\}$ shows its countability,
- an intuitive depiction of *Cantor enumeration* of the set of pairs of natural numbers is shown in Table 1.1. This intuition can be explicitly expressed by the formula $\pi(x, y) = (x + y) \cdot (x + y + 1)/2 + 1$, where (x, y) is an enumerated pair of natural numbers.

The Cantor enumeration of pairs of natural numbers reveals some interesting characteristics. We note that any rational number can be represented as a fraction of two integer numbers. Any non-negative fraction is formed by taking a column's label as the

https://doi.org/10.1515/9783112223123-001

Table 1.1: Cantor enumeration of the set of pairs of integer numbers. An intuitive depiction is shown below.

	0	**1**	**2**	**3**	**4**	**5**	**...**
0	0	1	3	6	10	15	...
1	2	4	7	11	16	22	...
2	5	8	12	17	23	...	...
3	9	13	18	24	...	...	...
4	14	19	25	24	...	...	...
5	20	26	...	...	...	...	...
...	...	...	...	...	...	...	...

numerator and treating a label of a row (except the first one) as its denominator. In this enumeration, any fraction appears once in irreducible form and infinitely many times in reducible form. Cantor enumeration of pairs of natural numbers could be adapted to an enumeration of non-negative rational numbers by skipping reducible fractions. On the other hand, since the set of rational numbers is a (proper) subset of pairs of natural numbers, the countability of the set of non-negative rational numbers is a direct result of the countability of the set of pairs of natural numbers. Finally, we can apply the way of enumeration of the set of integer numbers to all rational numbers.

Let us consider the unit interval of real numbers. Any member of this set could be represented in the form of infinite expansion. Assume that all real numbers are arranged in the following sequence:

$$\begin{array}{c} b_{00}\, b_{01}\, b_{02}\, b_{03}\, b_{04}\, b_{05}\, b_{06}\, b_{07}\, b_{08}\, b_{09} \dots \\ b_{10}\, b_{11}\, b_{12}\, b_{13}\, b_{14}\, b_{15}\, b_{16}\, b_{17}\, b_{18}\, b_{19} \dots \\ b_{20}\, b_{21}\, b_{22}\, b_{23}\, b_{24}\, b_{25}\, b_{26}\, b_{27}\, b_{28}\, b_{29} \dots \\ b_{30}\, b_{31}\, b_{32}\, b_{33}\, b_{34}\, b_{35}\, b_{36}\, b_{37}\, b_{38}\, b_{39} \dots \\ \dots \end{array}$$

The following real number of the unit interval does not appear in this sequence, where $\sim b$ is the binary digit opposite to the digit b, that is $\sim b = 0$ if and only if $b \neq 0$:

$$\sim b_{00} \sim b_{11} \sim b_{22} \sim b_{33} \sim b_{44} \sim b_{55} \dots$$

Therefore, the assumption that all real numbers of the unit interval could be arranged in a sequence is false.

Example 1.2. Let us consider the set $\Sigma = \{0,1\}$ of binary digits and the set $\Sigma^* = \{0,1\}^*$ of all binary strings of finite length. Binary strings can be put (arranged) in a so called *canonical order*, that is:

- shorter strings precede longer ones, and
- given two strings of the same length, this string comes first, which have 0 at the leftmost position that is different in both strings.

Denoting the *empty string* (of length 0) by ε we get the following sequence of binary strings in canonical order:

$$\Sigma^* = \{\varepsilon, 0, 1, 00, 01, 10, 11, 000, 001, 010, 011, 100, 101, 110, 111, 0000, 0001, \ldots\}$$

Canonical order of binary strings creates enumeration, which assigns natural numbers to consecutive strings of this family. The existence of such enumeration directly proves that the set of binary strings is countable.

Elementary operations performed on sets are: *taking subset, union, intersection, complement of a subset* and *Cartesian product.*

Let us recall that *Cartesian product* $X_1 \times X_2 \times \cdots \times X_n$ of sets $X_1, X_2, \ldots, X_n$ is the set of so called *n-tuples* $\{(x_1, x_2, \ldots, x_n) : x_1 \in X_1, x_2 \in X_2, \ldots, x_n \in X_n\}$.

Note that finite sets and countable sets are closed under the above operations. For instance, the result of the complement of a subset of a countable set is also countable.

1.2 Relations

Relations play a fundamental role in this book. They are exploited in interpretation of key ideas and concepts such as graphs, grammars, productions and derivations in grammars, transitions and computations of automata and machines. Relations are summoned up in this section while more specific relations concerning concepts covered in the book are brought up and discussed in consecutive sections.

1.2.1 Multi place relations

Multi place relations are also called multidimensional relations or n-ary relations. A relation is a subset of a given set of objects, usually a subset of the Cartesian product of a number of sets. In this book, relations are exclusively identified with Cartesian products of a number of nonempty sets. Elements of a relation are called tuples (or n-tuples in the case of elements of the Cartesian product composed of n sets). In particular, the n-tuples are referred to as follows: single (1-tuple), pair (2-tuple), triple, quadruple, quintuple, sextuple, etc. A multi-place relation defined in the Cartesian product of n sets is named an *n-place relation* or *n-ary relation.*

Definition 1.1. Any subset R of the Cartesian product $X_1 \times X_2 \times \cdots \times X_n$ of nonempty sets $X_1, X_2, \ldots, X_n$, $R \subset X_1 \times X_2 \times \cdots \times X_n$, is an n-ary relation (n-place or n-dimensional relation) over sets $X_1, X_2, \ldots, X_n$.

A tuple $(x_1, x_2, \ldots, x_n)$ where $x_1 \in X_1, x_2 \in X_2, \ldots, x_n \in X_n$, is said to be an element of the relation R or to satisfy the relation R if and only if $(x_1, x_2, \ldots, x_n) \in R$. Inclusion of the tuple $(x_1, x_2, \ldots, x_n) \in R$ into the relation R is denoted equivalently as $R(x_1, x_2, \ldots, x_n)$. We also say that elements $x_1, x_2, \dots, x_n$ are *R-related* (or – *related*, if this does lead to confusion). The tuple $x_1, x_2, \dots, x_n$ is not an element of the relation R or does not satisfy the relation R if and only if $(x_1, x_2, \dots, x_n) \notin R$. Exclusion of the tuple from the relation is also denoted as $\sim R(x_1, x_2, \dots, x_n)$. Here we also say that elements $x_1, x_2, \dots, x_n$ are not related.

Example 1.3. Let us consider the Cartesian product $Z \times Z \times Z \times Z$, where Z is the set of integer numbers. The relation $R \subset Z \times Z \times Z \times Z$, defined as the set of quadruples $R = \{(k, l, m, n) : k = l \cdot m + n, 0 < l, 0 \le n < |l|\}$, identifies the operation of integer division with dividend k, divisor l, quotient m and remainder n. For instance, $(4, 3, 1, 1) \in R$, $(-4, 3, -2, 2) \in R$, $(5, -3, -1, 2) \in R$ and $(-5, -3, 2, 1) \in R$ while $(2, 0, 2, 2) \notin R$ and $(-5, 3, -1, -2) \notin R$. Let us recall that remainder is a non-negative integer less than absolute value of divisor.

Note that an n-dimensional relation forms a system of $n + 1$ components: n sets $X_1, X_2, \dots, X_n$ and a subset of the Cartesian product $X_1 \times X_2 \times \dots \times X_n$. So then relations may differ either in one or more sets, or in a subset of the Cartesian product, or in both.

1.2.2 Two place relations

Two place relations are special cases of multi place relations. Any subset ρ of the Cartesian product $X \times Y$ of nonempty sets X and Y, $\rho \subset X \times Y$, is a two place relation. A pair (x,y) where $x \in X$ and $y \in Y$, is an element of the relation ρ if and only if $(x,y) \in \rho$. Inclusion of the pair (x,y) into the relation ρ is equivalently denoted as $\rho(x,y)$ or $x \; \rho \; y$. We say that such elements x, y are *ρ-related* (or *related*, for short). The pair (x,y) is not an element of the relation ρ or it does not satisfy the relation if and only if $(x,y) \notin \rho$. Exclusion of the pair (x,y) from the relation ρ is denoted $\sim \rho(x,y)$ or $\sim x \; \rho \; y$. We also say that such elements x, y are not *ρ-related.*

Definition 1.2. For a given two place relation $\rho \subset X \times Y$ we introduce the following terminology:

- the set X is *the domain* of the relation ρ,
- the relation is *defined* for an element $x \in X$ if and only if there exists an element $y \in Y$ such that $\rho(x,y)$. If for given element $x \in X$, an element $y \in Y$ such that $\rho(x,y)$ does not exist, then the relation ρ is *undefined* for that element $x \in X$,
- the set of all elements $x \in X$, for which the relation is defined, is alternatively called a domain of the relation. However, in this book, the whole set X is considered to be the domain of a two-place relation,
- the set Y is the *codomain* of the relation ρ,
- the set of all $y \in Y$, such that for a given y there exists $x \in X$ related to y, that is, $\rho(x,y)$, is the *range* of a relation,
- an element x is called the argument and an element y is called the value of the pair (x,y),
- a relation is:

- *total (left-total)* if and only if for all $x \in X$ there exists $y \in Y$ such that $\rho(x,y)$,
- *surjective (right-total)* if and only if for all $y \in Y$ there exists $x \in X$ such that $\rho(x,y)$,
- *injective (left-unique)* if and only if for all $x_1, x_2 \in X$ and $y \in Y$ it holds that if $\rho(x_1,y)$ and $\rho(x_2,y)$ then $x_1 = x_2$. Injective relation is a mapping (function),
- a *function* if and only if the relation is injective. Injective and total relation is a total function,
- *bijective* if and only if is left-total, surjective and injective. Bijective relation is also called *1-to-1 relation*. Obviously, bijective relation is a (1-to-1) mapping also called bijection.

Example 1.4. Let us define a relation $L \subset \Sigma^* \times N$, where $\Sigma = \{0,1\}$ denotes the set of binary digits, Σ^* denotes the set of binary strings (see Example 1.2) and N denotes the set of natural numbers, that is, $N = 0,1,2,3,\ldots$. The relation includes pairs of a string and its length: for any string $s \in \Sigma^*$ and for any natural number $n \in N$, $(s,n) \in L$, if and only if $|s| = n$, where $|s|$ denotes length of the string s, that is, the number of symbols in the string s. Note that L is left-total and right-total (surjective) but not injective. This observation is straightforward.

Example 1.5. Two-place relations with both countable domain and codomain can be represented in the form of two-dimensional tables. The rows of the table are labeled by the domain members, while the codomain members label the columns. The entries of this table are equal to 1 where the corresponding pair of row and column labels belong to the relation or is equal to 0 where the pair does not belong to the relation. The relation L defined in Example 1.4 is shown in Table 1.2.

Note that all binary strings can be enumerated according to canonical order. Consecutive strings are denoted by s_i, where indices enumerate strings, as is shown in the first column of Table 1.2. For instance, $s_0 = \varepsilon$, $s_1 = 0$, $s_2 = 1$, $s_3 = 00$ etc. This enumeration assigns to any string of length l an index not less than $2^l - 1$ and not greater than $2^{l+1} - 2$.

1.2.3 Binary relations

A particular case of two-place relations are relations with the same domain and codomain. In this book, such relations are called binary relations. In other words, a relation $\rho \in X \times X$ over X, where X is a nonempty set, is a binary relation.

We distinguish several important classes of binary relations where their taxonomy takes into consideration the following fundamental properties:
- *reflexivity:* $(\forall x \in X)\ x\rho x$
- *irreflexivity:* $(\forall x \in X) \sim x\rho x$
- *symmetry:* $(\forall x,y \in X)\ x\rho y \Rightarrow y\rho x$
- *antisymmetry:* $(\forall x,y \in X)\ x\rho y \Rightarrow \sim y\rho x$

Table 1.2: Representation of the relation L defined in Example 1.4 shown below (note: 0s are dropped for the sake of readability).

		0	1	2	3	4	...
s_0	ε	1					...
s_1	0		1				...
s_2	1		1				...
s_3	00			1			...
s_4	01			1			...
s_5	10			1			...
s_6	11			1			...
s_7	000				1		...
s_8	001				1		...
s_9	010				1		...
s_{10}	011				1		...
s_{11}	100				1		...
s_{12}	101				1		...
s_{13}	110				1		...
s_{14}	111				1		...
s_{15}	0000					1	...
s_{16}	0001					1	...
s_{17}	0010					1	...
...	...	...	...	...	...	...	...

- *asymmetry:* $(\forall x, y \in X)\ x\rho y$ and $y\rho x \Rightarrow x = y$
- *transitivity:* $(\forall x, y, z \in X)\ x\rho y$ and $y\rho z \Rightarrow x\rho z$
- *totality:* $(\forall x, y \in X)\ x\rho y$ or $y\rho x$
- *partial order:* a relation which is reflexive, asymmetric and transitive
- *total order:* a relation which is asymmetric, transitive and total

Example 1.6. The relation $\leq \in Z \times Z$ (*less or equal*) over the set Z of integers is reflexive, asymmetric, transitive and total, i. e. it is reflexive and possesses the property of total order. The relation $< \in Z \times Z$ (*less than*) over the set Z of integers is irreflexive, antisymmetric, and transitive. The relation *is a divisor of* over the set of positive natural numbers is reflexive, asymmetric and transitive, that is, it is partial order.

1.2.4 Equivalence relations

Equivalence relations formally exhibit the similarity of objects of a given set. In fact, equivalence relations identify groups of elements that are not distinguishable with regard to some aspects or properties.

Definition 1.3. Equivalence relation over a set X of objects is a binary relation that is reflexive, symmetric and transitive.

For instance, graph isomorphism is an equivalence relation defined in the set of graphs (it is elementary to prove reflexivity, symmetry and transitivity of graph isomorphism). Indeed, isomorphic graphs have the same structure and may only have different labels of vertices. Such differences are unimportant from the point of view of the features (properties).

Definition 1.4. A subset A of the set X is an equivalence class of an equivalence relation $\rho \in X \times X$ if and only if

$$(\forall x, y \in A)x\rho y \wedge (\forall x \in A)(\forall y \notin A) \sim x\rho y$$

For a given $a \in X$ and a given equivalence relation ρ, the set of all members of X related to a is an equivalence class denoted $[a]_\rho$.

It is straightforward to note that equivalence classes (of an equivalence relation) *split* the set X, that is, they are pairwise disjoint and every member of X belongs to one of them.

An equivalence relation can be outlined by defining its equivalence classes.

Example 1.7. Let us consider the set $\{0,1\}^*$ of binary strings and the binary relation over this set which relates any two strings of the same length. This relation is reflexive, symmetric and transitive, that is, it is an equivalence relation. The relation is illustrated in Table 1.3. Due to the canonical ordering of binary strings, the equivalence classes come as blocks of 1's shown in this Table.

Table 1.3: Representation of the relation defined in Example 1.7.

	ε	0	1	00	01	10	11	000	001	...
ε	1									...
0		1	1							...
1		1	1							...
00				1	1	1	1			...
01				1	1	1	1			...
10				1	1	1	1			...
11				1	1	1	1			...
000								1	1	...
001								1	1	...
...	...	...	...	...	...	...	...	...	...	...

1.2.5 Closure of relations

A question about a relation that includes arbitrarily given pairs of elements and is desired to exhibit given properties invokes the notion of closure of relation.

Definition 1.5. Let $R \subset X \times X$ is a binary relation and P is a set of properties. Closure of the relation R with respect to a set P of properties, P-closure of the relation for short, is the smallest binary relation R^P, which includes the relation R and posses all properties of P.

The reflexive closure $R^{\{r\}}$ of a binary relation $R \subset X \times X$ requires that every pair (x, x), $x \in X$ is $R^{\{r\}}$, that is, $R^{\{r\}}$ comes as the union $R^{\{r\}} = R \cup \{(x, x) : x \in X\}$. In the corresponding tabular representation, the reflexive closure sets to 1 all entries on the main diagonal.

Similarly, the symmetric closure necessitates the inclusion of any pair (x, y), $x, y \in X$ such that $(y, x) \in R : R^{\{s\}} = R \cup \{(x, y) : (y, x) \in R\}$. Concerning the tabular representation, the symmetric closure implies that the table is made symmetric with regard to the main diagonal; that is, zero's entries must be set to one if they are symmetric with respect to one's entries.

Example 1.8. Let us consider a binary relation over the set of all binary strings, $R \subset \{0,1\}^* \times \{0,1\}^*$. The relation includes every pair of two successive strings in canonical order. Find the closure $R^{\{r,t\}}$ of R concerning both reflexivity and transitivity.

Solution. The relation R is shown in Table 1.4 while the closure $R^{\{r,t\}}$ is outlined in Table 1.5. To prove that $R^{\{r,t\}}$ is the closure of the relation R with respect to the set of reflexivity and transitivity, let us notice that $R^{\{r,t\}}$ is reflexive (possesses 1's on the main diagonal). It is transitive, as well. Note that a pair $(s_i, s_j) \in R^{\{r,t\}}$ if and only if $i \le j$. Having $(s_i, s_j) \in R^{\{r,t\}}$ and $(s_j, s_k) \in R^{\{r,t\}}$ we get $i \le k$, since $i \le j$ and $j \le k$. Therefore $(s_i, s_k) \in R^{\{r,t\}}$.

Table 1.4: Representation of the relation R defined in Example 1.8 (as before 0s are not shown here).

		s_0	s_1	s_2	s_3	s_4	s_5	s_6	s_7	...
		ε	0	1	00	01	10	11	000	...
s_0	ε		1							...
s_1	0			1						...
s_2	1				1					...
s_3	00					1				...
s_4	01						1			...
s_5	10							1		...
s_6	11								1	...
s_7	000									...
...	...	...	...	...	...	...	...	...	...	...

To complete the proof it is enough to show that $R^{\{r,t\}}$ is the smallest reflexive and transitive relation including the relation R. Assume that this is not satisfied. As a result of this assumption, there exists a reflexive and transitive relation R' which includes R and does not include a pair (s_i, s_{i+k}), $i \ge 0$, $k \ge 0$. If $k = 0$ then R' cannot be reflexive.

Table 1.5: Representation of the relation $R^{\{r,t\}}$ defined in Example 1.8 (0s are not shown here).

		s_0	s_1	s_2	s_3	s_4	s_5	s_6	s_7	...
		ε	0	1	00	01	10	11	000	...
s_0	ε	1	1	1	1	1	1	1	1	...
s_1	0		1	1	1	1	1	1	1	...
s_2	1			1	1	1	1	1	1	...
s_3	00				1	1	1	1	1	...
s_4	01					1	1	1	1	...
s_5	10						1	1	1	...
s_6	11							1	1	...
s_7	000								1	...
...	...	...	...	...	...	...	...	...	...	...

If $k = 1$ then $R \not\subset R'$. Then we have $(s_i, s_{i+1}) \in R$, $(s_{i+1}, s_{i+2}) \in R, \dots, (s_{i+k-1}, s_{i+k}) \in R$. Subsequently, applying the property of transitivity k times we get that (s_i, s_{i+k}) should be included into transitive closure (being very formal, an inductive proof would be applied, cf. Section 1.3). Afterward the assumption that the relation $R^{\{r,t\}}$ is not the smallest one is false. This ends the proof.

1.3 Mathematical induction

Mathematical induction is a generic tool to prove properties involving natural numbers. It is used to establish that a property is satisfied for all natural numbers based on its satisfaction with the first natural number (0 or 1). Then it is proved that satisfaction for any natural number implies satisfaction for the next one.

The simplest and the most common formulation of mathematical induction is given in Definition 1.6. It can be applied in solving simple problems based on the whole set of natural numbers.

Definition 1.6. Let assume that W is a property defined in the set N of natural numbers such that:

- $W(0)$
  the basis of induction (0 satisfies the property W),
- $(\forall n \in N) W(n) \Rightarrow W(n+1)$
  the inductive step (for any natural number n the following holds: if n satisfies the property W, then so does $n+1$)

then all natural numbers satisfy the property W.

An equivalent formulation of mathematical induction given in Definition 1.7 allows us to use it in problems, in which a proof of the strictly consecutive law $W(n) \Rightarrow W(n+1)$ is difficult to be employed. However, it is easier to prove the corresponding law: if the property is satisfied for natural numbers smaller than or equal to n then so it is satisfied for $n + 1$.

Definition 1.7. Let assume that W is a property defined in the set N of natural numbers such that:

- $W(0)$
  the basis of induction (0 satisfies the property W),
- $(\forall n \in N)((\forall k = 1, 2, \ldots, n)W(k)) \Rightarrow W(n + 1)$
  the inductive step (for any natural number n it holds that if every $k = 0, 1, \ldots, n$ satisfies the property W, then $n + 1$ satisfies W)

then all natural numbers satisfy the property W.

The assumption in the inductive step, that is, that a property holds for a given natural number n or it holds for all numbers not greater than n, is also called the inductive hypothesis. Based on the assumption that the inductive hypothesis is true, the satisfaction of the property for $n + 1$ is drawn in the inductive step.

The principle of mathematical induction can be used for proving problems related to any countable set. Suppose a countable set is enumerated by making use of natural numbers. In that case, the problem can be related to indices, which are natural numbers rather than original members of the countable set. For instance, the principle of mathematical induction can be used in proofs of properties of a set of natural numbers greater or equal to some given number n_0, of the set of prime numbers, of the set of odd integers, of the set of rational numbers, of the set of all binary strings of even length, of graphs, etc. Inductive proofs involving both the above formulations of mathematical induction as well as different countable sets will be presented in consecutive sections.

1.4 Graphs and trees

Graphs are essential mathematical structures used to model problems in theory and practice. Trees are a particular type of graphs. In this book, graphs are mostly utilized for modeling automata and trees. They are also employed to represent computations realized by automata and to illustrate derivations of words in context-free grammars.

1.4.1 Graphs

Definition 1.8. A graph G is a system of two components $G = (V, E)$ where $V = \{v_1, v_2, \ldots, v_n\}$ is a finite set of vertices (vertices are also called nodes) and $E \subset V \times V$ is a binary relation over the set of vertices. Elements of $E = \{e_1, e_2, \ldots, e_m\}$ are called edges. Edges connect vertices. Sets of vertices and edges of a graph G are usually denoted

$V(G)$ and $E(G)$ or V and E when there is no danger of confusion. A graph $G = (V, E)$ is undirected if and only if the relation E is symmetric. Otherwise, a graph is directed.

In directed graphs, an edge connecting a vertex v_i to a vertex v_j is denoted as (v_i, v_j). In undirected graphs an edge connecting a vertex v_i and a vertex v_j is denoted $\{v_i, v_j\}$ and can be interpreted as the pair of directed edges (v_i, v_j) and (v_j, v_i).

Graphs are commonly represented in a graphical form, as adjacency matrices or as adjacency lists.

Rows and columns of an adjacency matrix are indexed by vertices. Entries of the matrix are equal to 0 or 1. An entry is equal to 0 if and only if there is no edge connecting vertices corresponding to labels of the row and the column.

Two types of adjacency lists are constructed for every vertex. For a given vertex v_j, a list *before* includes all vertices connected to v_j (such that there is an edge going from the given vertex to v_j). For a given vertex v_j, a list *after* includes all vertices to which v_j is connected (such that there is an edge going from v_j to the given vertex). Note that in undirected graphs for each vertex lists *before* and *after* are identical.

The graphical representation of adjacency matrices and adjacency lists of a directed graph and an undirected graph are shown in Figure 1.1.

a)

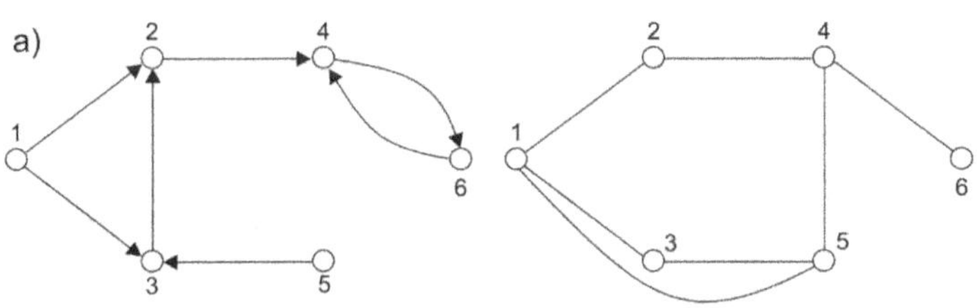

b)

	1	2	3	4	5	6
1	0	1	1	0	0	0
2	0	0	0	1	0	0
3	0	1	0	0	0	0
4	0	0	0	0	0	1
5	0	0	1	0	0	0
6	0	0	0	1	0	0

	1	2	3	4	5	6
1	0	1	1	0	1	0
2	1	0	0	1	0	0
3	1	0	0	0	1	0
4	0	1	0	0	1	1
5	1	0	1	1	0	0
6	0	0	0	1	0	0

c)

		1	2	3
1	3	2	4	
1	5	3	2	
2	6	4	6	
		5	3	
	4	6	4	

1	2	3	5
2	1	4	
3	1	5	
4	2	5	6
5	1	3	4
6	4		

Figure 1.1: Graphs and their representation: a) examples of directed and undirected graphs, b) adjacency matrices, c) adjacency lists (notice that lists *before* were dropped in case of the undirected graph).

A *path* in a graph $G = (V, E)$ is a sequence of vertices $v_{i_1}, v_{i_2}, \ldots, v_{i_k}$ that are pairwise different and such that $(v_{i_{j-1}}, v_{i_j}) \in E$ for $j = 2, 3, \ldots, k$ (in case of undirected graphs $\{v_{i_{j-1}}, v_{i_j}\}$ is considered instead of $(v_{i_{j-1}}, v_{i_j})$).

A path with $v_{i_1} = v_{i_k}$ is a *cycle*. A graph with no cycle is called *acyclic*.

A graph is *connected* if and only if there is a path between any two vertices.

1.4.2 Trees

In graph theory a *tree* T is defined as a connected acyclic *graph* G. In this book, we employ an inductive definition of a tree.

Definition 1.9. A tree is an undirected connected graph G having the following properties:

- the graph having one vertex $T = (V = \{v\}, E = \emptyset)$ forms a tree with the root v,
- if $T_1 = (V_1, E_1)$, $T_2 = (V_2, E_2), \ldots, T_k = (V_k, E_k)$ form trees with roots $v_1, v_2, \ldots, v_k$, respectively, and a vertex $v \notin V_1 \cup V_2 \cup \cdots \cup V_k$ then $T = (V_1 \cup V_2 \cup \cdots \cup V_k \cup \{v\}, E_1 \cup E_2 \cup \cdots \cup E_k \cup \{\{v, v_1\}, \{v, v_2\}, \ldots, \{v, v_k\}\})$ is the tree with the root v. Vertices $v_1, v_2, \ldots, v_k$ are children of the vertex v and v is their parent. $T_1, T_2, \ldots, T_k$ are subtrees of the tree T,
- a graph is a tree if and only if it is constructed employing the two above rules for finite number of times.

Vertices with no child are *leaves*. The set of all leaves of a tree is called *crop* of the tree.

A tree that is built using this definition is an undirected acyclic graph. The definition could be easily adapted for directed trees with edges connecting parents to children.

A tree T is called *k-tree* if and only if every node has no more than k children and there is a node with k children. The simplest trees are *2-trees*, also called *binary trees*.

Definition 1.10. Height of a tree is defined as follows:

- height of the tree $T = (V = \{v\}, E = \emptyset)$ is equal to 0,
- if $T_1, T_2, \ldots, T_k$ are trees of height $h_1, h_2, \ldots, h_k$, respectively, then the tree T with subtrees $T_1, T_2, \ldots, T_k$ has height equal to $1 + \max\{h_1, h_2, \ldots, h_k\}$.

Note that the height of a tree is equal to the length of the longest path from the root to a leaf.

Now we can prove the following important property of trees.

Lemma 1.1. *A k-tree of height h has no more than k^h leaves.*

Proof. By induction:

- a k-tree $T = (V = \{v\},\ E = \emptyset)$ of one vertex has 1 leaf, its height is equal to 0. Of course, $1 \le k^0$,

- if $T_1, T_2, \ldots, T_l, l \le k$, are k-trees of height not greater than h and – by inductive assumption – they have $N_1 \le k^h, N_2 \le k^h, \ldots, N_l \le k^h$ leaves, respectively, the tree T with subtrees $T_1, T_2, \ldots, T_l$ of height $h+1$ has $N_1 + N_2 + \cdots + N_l \le l \cdot k^h \le k \cdot k^h = k^{h+1}$ leaves,
- since the basis (the first condition) and the inductive step (the second condition) hold, then the lemma has been proved. □

1.5 Languages

Languages are the main subjects of this study. Let us recall that languages are subsets of the set of all words over some finite alphabet. An alphabet is a finite set of symbols, while a word is a finite sequence (finite string) of symbols of this alphabet. This description of languages has little in common with a typical understanding of languages such as *natural languages*: English, Spanish, Polish, Japanese, or even such as programming languages. As regarded here, languages are objects of set theory rather than natural languages or programming languages. However, the discussion in this book and conclusions drawn here are fundamental to various areas of research and practice. Such areas include design and implementation of compilers, construction of algorithms, analysis of the complexity of algorithms, and characterization of the space of problems from a perspective of the complexity of algorithms used to solve them.

1.5.1 Basic definitions

Definition 1.11. *Alphabet* is a finite set of symbols: $\Sigma = \{a^{(1)}, a^{(2)}, \ldots, a^{(\rho)}\}$. Symbols of an alphabet are also called *letters*.

Word over an alphabet Σ is a finite sequence $w = a_1\, a_2\, \ldots\, a_n$ of letters. Length of a word is equal to the number of symbols of the word and is denoted $|w|$. Words are also called *strings*.

The set of all words over alphabet Σ is denoted as Σ^*.

Note:

- symbols in a word are denoted using subscripts while symbols of an alphabet are indicated with superscripts in brackets. This is just to differentiate symbols of an alphabet from symbols of a word,
- the set Σ^* of all words over alphabet $\Sigma = \{a^{(1)}, a^{(2)}, \ldots, a^{(\rho)}\}$ is infinite and countable. Words over alphabet Σ can be arranged in the *canonical order*: longer words follow shorter ones and words of the same length are arranged alphabetically assuming order of letters in an alphabet consistent with their indexes, that is

$$\Sigma^* = \{\varepsilon, a^{(1)}, a^{(2)}, \ldots, a^{(\rho)}, a^{(1)}a^{(1)}, a^{(1)}a^{(2)}, \ldots, a^{(1)}a^{(\rho)},$$
$$a^{(2)}a^{(1)}, a^{(2)}a^{(2)}, \ldots, a^{(\rho)}a^{(\rho)}, a^{(1)}a^{(1)}a^{(1)}, \ldots\}$$

- words can be enumerated consistently with the canonical order, that is:
  - the empty word $w_0 = \varepsilon$ (of length 0) gets the number 0,
  - words $w_1 = a^{(1)}, \dots, w_\rho = a^{(\rho)}$ (of length 1) get numbers 1 to ρ,
  - words of length 2 get numbers $\rho + 1$ to $\rho + \rho^2$,
  - words of length 3 get numbers $\rho + \rho^2 + 1$ to $\rho + \rho^2 + \rho^3$,
  - words of length 4 get numbers $\rho + \rho^2 + \rho^3 + 1$ to $\rho + \rho^2 + \rho^3 + \rho^4$,
  - etc.

Example 1.9. Let us consider the Latin alphabet $\{a; b, c, \dots, z\}$ consisting of 26 letters. Examples of words enumerated in canonical order are: $w_0 = \varepsilon$, $w_1 = a$, $w_2 = b$, $w_{23} = w$, $w_{24} = x$, $w_{27} = aa$, $w_{702} = zz$, $w_{703} = aaa$ and so on.

Definition 1.12. A language L over an alphabet Σ is a subset of the set of all words Σ^* over this alphabet.

Example 1.10. Examples of languages over the Latin alphabet:
- the empty language $\emptyset$,
- the language of words including only letter a: $\{a, aa, aaa, aaaa, \dots\}$,
- the language of one letter words: $\{a, b, c. \dots, z\}$,
- the language of all correct English words,
- the language of palindromes (a palindrome is a word which is identical to its reverse, e. g. *abcdcba*, *aaabbaaa*.

Note: a language is *finite* (that is, it has a finite number of words) if and only if there is some constant such that the length of any word in this language is not greater than this constant. Indeed, if the length of words of the language is bounded by a constant n, then the number of all words of this language of length not greater than n is finite and not greater than $1 + \rho + \rho^2 + \dots + \rho^n$. So then, the language is finite. On the other hand, if a language is finite, there is the longest word (more than one word may have the same length). The length of the longest word is the constant bounding length of all words of this language.

1.5.2 Operations on languages

In this section, operations on languages are discussed. Languages themselves are sets, so then intuitively appealing set operations (taking subset, union, intersection, complement) are considered here. More operations are discussed below and in Part III.

Concatenation is an important operation realized on languages. This operation relies on joining (concatenating) words. Let us consider two words, say $u = a_1a_2 \dots a_k$ and $v = b_1b_2 \dots b_l$. The concatenation of these two words is the word $w = a_1a_2 \dots a_kb_1b_2 \dots b_l$ denoted as $w = uv$ or $w = u \circ v$.

Concatenation of two languages L_1 and L_2 is the language $L_1 \circ L_2 = \{uv : u \in L_1, v \in L_2\}$. Concatenation of two languages over different alphabets produces a language over the union of both alphabets.

Concatenation of two languages could be easily generalized to concatenation of a finite sequence of languages. Concatenation of languages $L_1, L_2, \ldots, L_n$ is the language $(\ldots((L_1 \circ L_2) \circ L_3)\ldots) \circ L_n$.

Let us notice that concatenation is an associative operation both on words and on languages. The proof of this is straightforward.

A special focus is placed on the concatenation of the same language. *n-concatenation* of a language L, denoted by L^n, is defined inductively in the following way

$$L^0 = \{\varepsilon\}$$
$$L^{n+1} = L^n \circ L$$

This operation gives rise to the so-called *Kleene closure* or *star closure*. The Kleene closure of a language L is the language L^*:

$$L^* = \bigcup_{k=0}^{\infty} L^k$$

For the sake of simplicity, we use the symbol of *positive closure*:

$$L^+ = \bigcup_{k=1}^{\infty} L^k$$

Note that either $L^+ = L^*$ when $\varepsilon \in L$ or $L^+ = L^* - \{\varepsilon\}$ when $\varepsilon \notin L$.

Example 1.11. Let us consider the language $L = \{0, 00, 000\}$. Multi-concatenation of the langue L produces the following results:
- $L^0 = \{\varepsilon\}$,
- $L^1 = L^0 \circ L = \{\varepsilon\} \circ \{0, 00, 000\} = \{\varepsilon \circ 0, \varepsilon \circ 00, \varepsilon \circ 000\} = \{0, 00, 000\}$,
- $L^2 = L^1 \circ L = \{0, 00, 000\} \circ \{0, 00, 000\} = \{0 \circ 0, 0 \circ 00, 0 \circ 000, 00 \circ 0, 00 \circ 00, 00 \circ 000, 000 \circ 0, 000 \circ 00, 000 \circ 000\} = \{00, 000, 0000, 00000, 000000\}$,
- $L^3 = L^2 \circ L = \{000, 0000, 00000, 000000, 0000000, 00000000, 000000000\}$,
- etc.
- $L^* = \{\varepsilon, 0, 00, 000, 0000, 00000, \ldots\}$,
- $L^+ = \{0, 00, 000, 0000, 00000, \ldots\}$.

To simplify the notation, the following symbols are used:
- $a^n = \underbrace{aa\ldots a}_{n \text{ times}}$ – the word of n letters a,
- $a^* = \{a^0, a^1, a^2, a^3, \ldots\} = \{\varepsilon, a, aa, aaa, \ldots\}$. That is, a^* is the language $L = \{a\}^*$,

- $\sharp_a w$ denotes the number of letters a in a word w. For example, for $w = abacbb$, $\sharp_a w = 2$, $\sharp_b w = 3$, $\sharp_d w = 0$ where a, b, c, d are letters of the alphabet.

Now we consider two special binary relations defined on the set of all words Σ^* over a given alphabet. The first one, the right invariant relation, is independent on any language. The second one, called the relation induced by a language, is defined for any language L.

Definition 1.13. A relation $R \subset \Sigma^* \times \Sigma^*$ is right invariant if and only if

$$(\forall u, v \in \Sigma^*)(u\,R\,v \Rightarrow (\forall z \in \Sigma^*)uz\,R\,vz)$$

Definition 1.14. A relation $R_L \subset \Sigma^* \times \Sigma^*$ is induced by a language $L \subset \Sigma^*$ if and only if

$$(\forall u, v \in \Sigma^*)(u\,R_L\,v \Leftrightarrow ((\forall z \in \Sigma^*)uz \in L \Leftrightarrow vz \in L))$$

Remark 1.1. Relation R_L induced by a language L is right invariant relation.

Proof. This observation is valid since $u\,R_L\,v$ implies that for any $z \in \Sigma^*$, $uz\,R_L\,vz$. Indeed, $uz\,R_L\,vz$ if and only if $(\forall y \in \Sigma^*)(uz)y \in L \Leftrightarrow (vz)y \in L$. On the other hand, the concatenation of words is associative, so $((uz)y \in L \Leftrightarrow (vz)y \in L) \Leftrightarrow (u(zy) \in L \Leftrightarrow v(zy) \in L)$. Denoting zy by x we get $ux \in L \Leftrightarrow vx \in L$, which satisfies the definition of R_L. □

Remark 1.2. Relation R_L induced by a language L is an equivalence relation.

Proof. Reflexivity and symmetry of the relation R_L is obvious (is directly derived from the same properties of the equivalence relation). Transitivity of the R_L is derived from properties of the equivalence relation as well. That is, the relation R_L is transitive if and only if for any $u, v, w \in \Sigma^*$ the following implication holds: $u\,R_L\,v$ and $v\,R_L\,w$ implies $u\,R_L\,w$. From definition of R_L we have $u\,R_L\,v \Leftrightarrow ((\forall x \in \Sigma^*)ux \in L \Leftrightarrow vx \in L)$ and $v\,R_L\,w \Leftrightarrow ((\forall y \in \Sigma^*)uy \in L \Leftrightarrow vy \in L)\,R_L$. Assuming $x = y = z$ for any $z \in \Sigma^*$ we obtain $uz \in L \Leftrightarrow vz \in L$ and $vz \in L \Leftrightarrow wz \in L$ and, finally, $uz \in L \Leftrightarrow wz \in L$. This completes the proof. □

Remark 1.3. Any language L is a union of some equivalence classes of the relation R_L induced by L.

Proof. Notice that each equivalence class either is included in the language L or is disjoint with it and no equivalence class can be shared between the language L and its complement. This is because $u\,R_L\,v$ implies that $u \in L \Leftrightarrow v \in L$. □

Remark 1.4. It is easy to notice that right invariant relation may not satisfy properties of equivalence relation. For instance, by relating words of different lengths, we get a right invariant relation. However, this relation cannot be an equivalence relation since it is neither reflexive nor transitive. Since any relation induced by a language is an equivalence relation, then this particular relation (a right invariant) cannot be induced by any language.

1.6 Grammars

Grammars are essential tools used for describing, defining and analyzing languages. At first glance, grammars can be seen as tools describing syntactic rules that manage the analysis and processing of natural languages. However, a complete and precise description of natural languages is impossible due to their inherent complexity. Grammars of natural languages describe rules though the rules are neither accurate, nor complete, nor rigorous. Correctness rules of natural languages are often not strict and thus can be used more flexibly.

In contrast, the languages studied in this book are often less complex than natural languages and are always precisely defined. They can be called *formal languages* or *artificial languages*. The correctness rules of such languages are always strict. Any construct (word, text) can be exclusively qualified either as a correct structure of the language or as not belonging to this language. Grammars are important tools describing artificial languages. Certainly, grammars defining artificial languages are much less complex than grammars of natural language and have rules precisely defined.

Definition 1.15. A grammar is a system

$$G = (V, T, P, S)$$

where:
- V is a finite set of variables (or nonterminals), will also be called alphabet of variables (alphabet of nonterminals),
- T is a finite set of terminals, will be also called alphabet of terminals,
- P is a finite set of productions,
- S is a variable called initial (or starting) symbols of the grammar.

Production is a pair (α, β) of words over alphabets of nonterminals and terminals assuming that α is a nonempty word:

$$(\alpha, \beta) \in (V \cup T)^{+} \times (V \cup T)^{*}$$

The set of grammar productions can be interpreted as a finite binary relation in the set of all words over the union of alphabets of nonterminals and terminals. Productions can also be interpreted as rules that allow us to change words by replacing a part of it identical to the first element of production with the second element of this production. This interpretation carries out another symbol of the production (α, β), that is

$$\alpha \rightarrow \beta$$

Intuitively, straightforward interpretation of productions as rules of words exchange leads to a notion of direct derivation and derivation:

Definition 1.16. A direct derivation $\underset{G}{\rightarrow}$ in a grammar $G = (V, T, P, S)$ is a binary relation:

$$\underset{G}{\rightarrow} \subset (V \cup T)^* \times (V \cup T)^*$$

where: $(\eta, \zeta) \in \underset{G}{\rightarrow}$ if and only if $(\exists \alpha, \beta, \gamma, \delta \in (V \cup T)^*)$ such that $\eta = \gamma\alpha\delta$, $\zeta = \gamma\beta\delta$ and $\alpha \rightarrow \beta$ is a production. Note: the grammar name G will be omitted in the derivation symbol if this does not lead to confusion.

Definition 1.17. A derivation $\overset{*}{\underset{G}{\rightarrow}}$ in a grammar $G = (V, T, P, S)$ is the transitive closure of a direct derivation. Obviously, a derivation is a binary relation in the set of all words over the union of sets of nonterminals and terminals.

A derivation allows verifying whether a word can be derived from another one by utilizing a sequence of productions.

Terminals of a grammar could be seen as an alphabet of letters used for the construction of words of a language over this alphabet. Nonterminals represent more general concepts. Such concepts should be developed by applying relevant productions as far as a final word over the terminal alphabet is derived.

Example 1.12. Let us consider the grammar $G = (V, T, P, S)$ where:

$$\begin{array}{llccc} V & = \{S, A\} & & & \\ T & = \{0, 1\} & & & \\ P: & S & \rightarrow & 0 & (1) \\ & S & \rightarrow & A & (2) \\ & A & \rightarrow & A0 & (3) \\ & A & \rightarrow & A1 & (4) \\ & A & \rightarrow & 1 & (5) \end{array}$$

Note that several productions have the same left-hand side. Such productions are often arranged together in a single line with the same left-hand side. The corresponding right-hand sides are separated by vertical bars, say

$$\begin{array}{lllll} P: & S & \rightarrow & 0|A & (1), (2) \\ & A & \rightarrow & A0|A1|1 & (3), (4), (5) \end{array}$$

Finally, as was mentioned above, grammar is a tool to define a language, which is outlined in the following definition.

Definition 1.18. A language $L(G)$ generated by the grammar (or generated in the grammar) G is the set of words over the alphabet of terminals, which could be derived from starting symbol of the grammar:

$$L(G) = \{w \in T^* : S \overset{*}{\underset{G}{\rightarrow}} w\}$$

Example 1.13. Let us consider the grammar $G = (V, T, P, S)$ defined in the previous example. Observe that this grammar generates the language $L(G) = \{0, 1, 10, 11, 100, 101, 110, 111, 1000, 1001, 1010, 1011, \ldots\}$, that is, it generates binary numbers without non-significant zeros.[1]

If not stated explicitly, the following symbols will be used by default to indicate constructs of grammars:

$A, B, C, \ldots$	nonterminal symbols,
$a, b, c, \ldots, 0, 1, 2, \ldots, 9$	terminal symbols,
$Z, Y, X, \ldots$	terminal or nonterminal symbol,
$z, y, x, \ldots$	words over alphabet of terminals,
$\alpha, \beta, \gamma, \ldots$	sequences of terminals and nonterminals.

1.7 Problems

Problem 1.1. Determine the number of all reflexive relations in a set of n elements $\{a_1, a_2, \ldots, a_n\}$.

Solution. Let us represent relations as tables with rows and columns labeled with elements of the set. The entries of this table are equal to 1 if the row label is related to the column label, cf. Example 1.5

The reflexive relation must have all entries on the main diagonal being equal to 1. Any other entry of the table may be either 0 or to 1. The number of entries outside the main diagonal is equal to $n^2 - n$, see Table 1.6. These entries may have any value (0 or 1). Therefore, the number of different reflexive relations is equal to 2^{n^2-n}.

Table 1.6: A reflexive relation defined on n elements.

	a_1	a_2	a_3	...	a_n
a_1	1	?	?	...	?
a_2	?	1	?	...	?
a_3	?	?	1	...	?
...	...	...	...	...	...
a_n	?	?	?	...	1

Problem 1.2. Find the number of all symmetric relations in a set of n elements $\{a_1, a_2, \ldots, a_n\}$.

Solution. Answer: $2^{(n^2+n)/2}$

---

1 A digit 0 in a number is non-significant if it can be removed without affecting the value of the number.

Problem 1.3. A relation ρ relates two words over binary alphabet $\Sigma = \{0,1\}$ having the same numerical value. Check if the relation ρ is the right invariant one. Find its equivalence closure and equivalence classes of the closure.

Solution. Let us check if this is an equivalence relation. Of course, the property of having the same numerical value is reflexive, symmetric and transitive. So then, its equivalence closure keeps the relation unchanged. Also, equivalence classes of this relation could be easily listed:

$$A_0 = \{0, 00, 000, 0000, 00000, \ldots\}$$
$$A_1 = \{1, 01, 001, 0001, 00001, \ldots\}$$
$$A_2 = \{10, 010, 0010, 00010, 000010, \ldots\}$$
$$A_3 = \{11, 011, 0011, 00011, 000011, \ldots\}$$
$$A_4 = \{100, 0100, 00100, 000100, 0000100, \ldots\}, \quad \text{etc.}$$

As we see, the relation creates an infinite set of infinite equivalence classes.

Two related words have the same numerical value. If any binary string is attached to both words, then their numerical values will be equal. This observation fulfills the definition of right invariant relation.

Problem 1.4. A relation ρ relates two words over binary alphabet $\Sigma = \{0,1\}$ having a common digit. For instance: $001\,\rho\,10$, $1010\,\rho\,010101$, $\sim 00\,\rho\,111$. Check if ρ is an equivalence relation. If not, find its equivalence closure $\rho^{\{eq\}}$. Find equivalence classes of the equivalence relation.

Solution. The relation is not an equivalence relation:
- it is not reflexive: any word is related to itself except the empty word, which is not related to itself,
- it is symmetric since the property of having common letter is symmetric,
- it is not transitive, for instance $00\,\rho\,01 \wedge 01\,\rho\,11$ and $\sim 00\,\rho\,11$.

Let us find the equivalence closure of the relation ρ. First of all, we will find its reflexive closure and its transitive closure. Then, we will check if the union of both closures remains reflexive, symmetric, transitive and is the smallest relation possessing these properties and including the relation ρ:
- it is obvious that the relation $\rho^{\{r\}} = \rho \cup \{(\varepsilon, \varepsilon)\}$ closes ρ over reflexivity: $\rho^{\{r\}}$ includes ρ, is reflexive and is the smallest relation satisfying these two conditions,
- note that any word including both digits is related to any nonempty word. Since for $w = 01$ and $(\forall u, v \in \Sigma^+) u\,\rho\,w \wedge w\,\rho\,v$, then the pair (u, v) should be included into transitive closure $\rho^{\{t\}}$ of the relation ρ. Indeed, any two nonempty words are related in the transitive closure of the relation ρ. On the other hand, empty word is not ρ-related to any other word. Concluding, we have $\rho^{\{t\}} = \Sigma^+ \times \Sigma^+$,

– we claim that $\rho^{\{r\}} \cup \rho^{\{t\}}$ is the smallest equivalence relation including the relation ρ. In fact, the relation $\rho^{\{r\}} \cup \rho^{\{t\}}$ is reflexive, symmetric and transitive, so it is an equivalence relation. Furthermore, it is not possible to remove any pair from the relation $\rho^{\{r\}} \cup \rho^{\{t\}}$ without violation of closure conditions: removing the pair $(\varepsilon, \varepsilon)$ violates reflexivity, removing a pair of words having common digit violates inclusion of the original relation, while removing any pair of nonempty words not having a common digit violates the transitivity property. In conclusion, the relation:

$$\rho^{\{eq\}} = \rho^{\{r\}} \cup \rho^{\{t\}} = \{(u,v) : u = \varepsilon = v \vee u \neq \varepsilon \wedge v \neq \varepsilon\} = \{(\varepsilon,\varepsilon)\} \cup \Sigma^+ \times \Sigma^+$$

is the equivalence closure of the original relation ρ.

In addition, let us note that $\rho^{\{eq\}}$ results in two equivalence classes: $E_1 = \{\varepsilon\}$ and $E_2 = \Sigma^+$.

Problem 1.5. Given is an alphabet $\Sigma = \{a^{(1)}, a^{(2)}, \ldots, a^{(r)}\}$, $r > 2$. The set of letters appearing in a word $w \in \Sigma^*$ is called *support* of the word and is denoted as $\mathrm{supp}(w)$. For instance, $\mathrm{supp}(a^{(1)}a^{(1)}a^{(1)}a^{(2)}) = \{a^{(1)}, a^{(2)}\}$. Words over the alphabet Σ are ρ-related if and only if they have exactly two common letters, that is, $(\forall u, v \in \Sigma^*) u \,\rho\, v \Leftrightarrow |\mathrm{supp}(u) \cap \mathrm{supp}(v)| = 2$. For instance, $001 \,\rho\, 10$, $1010 \,\rho\, 011$, $\sim 1010 \,\rho\, 11$, $\sim \varepsilon \,\rho\, 101$. Check if ρ is an equivalence relation. If not, find its equivalence closure $\rho^{\{eq\}}$. Find equivalence classes of the equivalence relation.

Solution. The relation is symmetric but is neither reflexive, nor transitive, for instance $\sim X \rho X$ for $X \in \Sigma \cup \{\varepsilon\}$; for a_1, a_2 and a_3 pairwise different letters $a_1a_2a_3 \,\rho\, a_1a_2 \wedge a_1a_2 \,\rho\, a_1a_2a_3$ and $\sim a_1a_2a_3 \,\rho\, a_1a_2a_3$. The reflexive closure is $\rho^{\{r\}} = \rho \cup \{(u,u) : u \in \Sigma^*\}$ while the transitive closure is $\rho^t = \{(u,v) : |\mathrm{supp}(u) \cap \mathrm{supp}(v)| \geq 2\}$ and the equivalence closure is equal to $\rho^{\{eq\}} = \rho^r \cup \rho^t$.

Equivalent classes of this closure are:

- $\{\varepsilon\}$,
- $\{a^i\}$ such that $a \in \Sigma$ and $i \in \{1, 2, \ldots\}$, that is, all sets including a single word with one letter support,
- $\{w \in \Sigma^* : |\mathrm{supp}(w)| \geq 2\}$, that is, the set of all words having at least two different letters.

The reader can easily prove that the closures and equivalent classes are correctly figured out.

Part I: **Grammars and generating languages**

2 Regular expressions and regular languages

Regular languages form the simplest class of formal languages. They are inductively defined by regular expressions. Regular languages also outline simple algebraic structures in the set of all words over a given alphabet. Due to this simplicity, they can be easily distinguished and identified. Regular languages are generated by regular grammars as well and – what will be discussed in Chapter 7 – they are accepted by finite automata.

In this Chapter, fundamental properties of regular expressions and regular languages are studied. We cover a discussion on the following topics: operation on regular expressions, algebraic properties of the relation induced by regular languages (portrayed by so-called Myhill–Nerode lemma, which is a part of Myhill–Nerode theorem), the structure of words of regular languages (delineated by pumping lemma). The study is supplemented by a section focused on regular grammars. All these topics are revisited in Chapter 8.

2.1 Regular expressions and regular languages

2.1.1 Regular expressions

The definition of regular expressions is inductive, that is, basic regular expressions are defined explicitly, while more complex regular expressions could be designed according to given rules.

Definition 2.1. Regular expressions over an alphabet Σ are constructs defined as follows:
- $\mathbf{\Phi}$ is a regular expression,
- $\boldsymbol{\varepsilon}$ is a regular expression,
- for each a in Σ, **a** is a regular expression,
- if r and s are regular expressions then
  - $(r + s)$, the sum of regular expressions,
  - (rs), the concatenation of regular expressions, and
  - (r^*), the Kleene closure of the regular expression, called also star closure

  are regular expressions.

Regular expressions are strings of basic symbols connected with sum, concatenation and Kleene operator enclosed in brackets. Basic symbols are shown in boldface in order to distinguish between the symbols of basic regular expressions and letters of the alphabet. Normal fonts will be used instead of boldface if the meaning of a symbol is obvious from the context it is used in.

Example 2.1. The following strings are regular expressions over the alphabet $\Sigma = \{a, b\}$:
- Φ,
- $\boldsymbol{\varepsilon}$,

https://doi.org/10.1515/9783112223123-003

- $\mathbf{a}$,
- $\mathbf{b}$,
- $(\Phi + \mathbf{a})$,
- $(\mathbf{b} + \boldsymbol{\varepsilon})$,
- $((\mathbf{aa})(\mathbf{ba}))$,
- $(((\mathbf{aa})\mathbf{b})(\mathbf{b} + \mathbf{a}))$,
- $(\mathbf{a} + (\mathbf{a}((\mathbf{bb})^*)))$.

Definition 2.2. Regular expressions over an alphabet Σ generate languages:
- Φ generates the empty language $\emptyset$:
- $\boldsymbol{\varepsilon}$ generates the language $\{\varepsilon\}$, that is, the empty word sets up this language,
- for each a in Σ, **a** generates the language $\{a\}$, that is, the one letter word sets up such a language,
- if regular expressions r and s generate languages R and S then
  - $(r + s)$ generates the language $R \cup S$ (union of languages R and S),
  - (rs) generates the language $R \circ S$ (concatenation of languages R and S),
  - $(r)^*$ generates the language R^* (Kleene closure of the language R).

Example 2.2. The following languages are generated by regular expressions used in Example 2.1:
- $\emptyset$, the empty language,
- $\{\varepsilon\}$ – the language having only the empty word,
- $\{a\}$ – the language having only one word of a single letter,
- $\{b\}$,
- $(\emptyset \cup \{a\}) = \{a\}$,
- $(\{b\} \cup \{\varepsilon\}) = \{\varepsilon, b\}$,
- $((\{a\}\{a\})(\{b\}\{a\})) = \{aaba\}$,
- $((\{a\}\{a\}\{b\})(\{a\} \cup \{b\})) = \{aaba, aabb\}$,
- $(\{a\} \cup (\{a\}((\{b\}\{b\})^*))) = \{a, abb, abbbb, abbbbbb, \ldots\}$.

Remark 2.1. Regular expressions are finite constructs. In fact, they are words over the alphabet $\Sigma \cup \{\Phi, \varepsilon, +, \circ, ^*, (,)\}$, where Σ is an alphabet of a regular expression. On the other hand, languages generated by regular expressions can be infinite. An example of infinite language generated by the (finite) regular expression is given in Example 2.2, refer to the last case.

Remark 2.2. Regular expressions and formulas describing languages generated by regular expressions include a large number of brackets, what makes them hardly readable. On the other hand, algebraic operators, logic operators as well as set theoretic operators are assumed to have priorities. This assumption allows dropping most of the brackets of algebraic, logic as well as set-theoretic expressions. By analogy, the same simplification is assumed for regular expressions. It is assumed that the sum operator has the lowest priority, concatenation has higher priority and the Kleene operator has the highest priority. The same assumption is adapted for operators on languages. Union has the lowest priority, concatenation has higher priority and the Kleene closure is of the highest priority. We will drop any pair of brackets if it does not change the order of operators when priorities are applied. As an example, a simplified form of regular expressions of Example 2.1 is presented in Example 2.3.

Example 2.3. Simplified form of regular expressions given in Example 2.1 presented below:

- Φ,
- $\boldsymbol{\varepsilon}$,
- **a**,
- **b**,
- Φ+**a**,
- $\mathbf{b} + \boldsymbol{\varepsilon}$,
- **aaba**,
- $\mathbf{aab(b + a)}$,
- $\mathbf{a + a(bb)}^*$.

Remark 2.3. Regular expressions can be interpreted as strings of symbols. Therefore, a simplified form of a regular expression is not equal to its original form in terms of strings' equality. Yet, languages generated by both forms of a regular expression are identical. So then, both forms are considered to be equivalent. From now on, if not stated otherwise, equivalent regular expressions will be deemed to be equal.

2.1.2 Regular languages

Definition 2.3. Regular languages are those and only those generated by regular expressions.

Example 2.4. The set of binary natural numbers without non-significant zeros $L = \{0, 1, 10, 11, 100, 101, 110, 111, 1000, 1001, \ldots\}$ is a language over the binary alphabet. Justify that it is a regular language.

Solution. Let assume that $\Sigma = \{0, 1\}$ is the alphabet. The regular expression $0+1(0+1)^*$ is claimed to generate the language L. Indeed, the regular expression $0+1(0+1)^*$ generates the language $\{0\} \cup \{1\} \circ \{0, 1\}^*$ which includes either word 0 or words beginning with 1. This proves that words generated by the regular expression $0 + 1(0 + 1)^*$ belong to the language L. On the other hand, any word of this language is generated by the expression. Obviously, 0 is generated by the expression. Furthermore, any word of length n and beginning with 1 belongs to the set $\{1\} \circ \{0, 1\}^{n-1}$ which is a subset of the language $\{0\} \cup \{1\} \circ \{0, 1\}^*$.

Example 2.5. Justify that the set L of binary words having exactly one sequence of three 1's 111 is a regular language. E. g., 0000111 $\in L$, 1111 $\notin L$ (two sequences of 111 are included: the first three digits and the last three digits) 11011001110001 $\in L$, 0110110 $\notin L$ (no sequence 111 included).

Solution. The following regular expression generates the language L:

$$r = 0^*((\varepsilon + 1)100^*)^*111(00^*1(\varepsilon + 1))^*0^*$$

The expression can be split into three parts: the middle 111, the beginning $0^*((\varepsilon+1)100^*)^*$ and the ending $(00^*1(\varepsilon + 1))^*0^*$. The middle part generates required three 1s.

The first part 0^* of $0^*((\varepsilon + 1)100^*)^*$ generates a sequence of 0's. The second part generates a sequence of binary words having one or two 1's followed by a nonempty sequence of 0's. Notice that each single or double digit 1 is followed by at least one 0.

The ending part (suffix) is the reversed beginning part (prefix). So then words generated by the ending part are reversed versions of words generated by the beginning part.

Finally, a concatenation of any words generated by the beginning, middle and ending part of the regular expression, r create words having exactly one sequence of three 1s, that's why they belong to the language L.

On the other hand, any word of the language L can be split into three parts: the middle sequence of three 1's, the sequence of digits (possibly empty) followed by the middle part and the sequence of digits that follows the middle part.

The beginning part of every nonempty word ends with digit 0 and, if have 1's, any single digit 1 or pair of them must be separated by at least a single 0. So then, the beginning part can be split into leading 0's and then sequences of single or double digits 1 followed by at least one 0. This observation makes it clear that the beginning part of any word of the language L is generated by the first part $0^*((\varepsilon+1)100^*)^*$ of the expression r.

By analogy, it can be drawn that the ending part $(00^*1(\varepsilon + 1))^*0^*$ of the regular expressions r generates the ending part of any word of the language L.

In conclusion, any word of the language L is generated by the regular expression r. This completes the proof.

Remark 2.4. The following equalities hold (in terms of Remark 2.3).
1. $\emptyset + r = r + \emptyset = r$
2. $\emptyset r = r\emptyset = \emptyset$
3. $\varepsilon r = r\varepsilon = r$
4. $\varepsilon + r = r + \varepsilon$
5. $r + s = s + r$
6. $(r + s) + t = r + (s + t) = r + s + t$
7. $(rs)t = r(st) = rst$
8. $r(s + t) = rs + rt$
9. $(r + s)t = rt + st$

10. $(r^*)^* = r^*$
11. $(r^*s^*)^* = (r + s)^*$
12. $(r^* + s^*)^* = (r + s)^*$

2.1.3 The Myhill–Nerode lemma

The Myhill–Nerode theorem is the most essential (or fundamental) tool characterizing regular languages. The theorem is given and proved in Chapter 8. In this Chapter a limited version of the Myhill–Nerode theorem is formulated. It will be referred to as the Myhill–Nerode lemma. The proof of the lemma is a direct consequence of a proof of the Myhill–Nerode theorem. The lemma is not proved here since important topics used in the proof have not been discussed yet. The lemma is formulated here since it is a crucial tool used in the identification of regular languages. It will be used in this and other Chapters.

Lemma 2.1 (The Myhill–Nerode lemma). *A language L is regular if and only if the relation R_L induced by the language L has a finite number of equivalence classes.*

Example 2.6. Verify if the language L of binary numbers without non-significant 0's is regular (this language was defined in Example 1.13).

Solution. Let check if the following sets are equivalence classes of the relation R_L induced by the language:
- $E_1 = \{\varepsilon\}$,
- $E_2 = \{0\}$,
- $E_3 = \{1w : w \in \{0,1\}^*\}$, that is, the set of all binary words beginning with the digit 1,
- $E_4 = \{0,1\}^* - (E_1 \cup E_2 \cup E_3)$, that is, the set of all words not included in the previous sets.

To prove that these sets are equivalence classes is sufficient to show that:
- they partition the set $\{0,1\}^*$ of all binary words. Certainly, because sets E_1, E_2 and E_3 are pairwise disjoint, then the definition of the set E_4 guarantees that these four sets create a partition of $\{0,1\}^*$,
- any two words of any set are R_L-related:
  - this condition is obvious for sets E_1 and E_2,
  - any two words u and v of the set E_3 belong to the language L (both words do not have non-significant 0). Moreover, for any binary word z, words uz and vz also belong to the language L (none of them have non-significant 0),
  - any two words u and v of the set E_4 do not belong to the language L (both words begin with non-significant 0). Moreover, for any binary word z, words uz and vz also do not belong to the language L (both of them begin with non-significant 0's),
- any two words of two different sets are not R_L-related:

- the empty word ε:
  * is not related to 0: for $z = \varepsilon$ we get $\varepsilon z = \varepsilon\varepsilon = \varepsilon$ and $0z = 0\varepsilon = 0$. Of course, $\varepsilon \notin L$ and $0 \in L$,
  * is not related to any word $u \in E_3$: for $z = \varepsilon$ we get $\varepsilon z \notin L$ and $uz \in L$,
  * is not related to any word $u \in E_4$: for $z = 0$ we get $\varepsilon z \in L$ and $uz \notin L$,
- the word 0
  * is not related to any word $u \in E_3$: for $z = 0$ we get $uz = 00$, so then $0z \notin L$ and $uz \in L$,
  * is not related to any word $u \in E_4$: for $z = \varepsilon$ we get $uz = 0$, so then $0z \in L$ and $uz \notin L$,
- any word $u \in E_3$
  * is not related to any word $v \in E_4$: for $z = \varepsilon$ we get $uz = u$, so then $uz \in L$ and $vz \notin L$.

Finally, by the Myhill–Nerode lemma, since the number of equivalence classes of the relation R_L induced by the language L is finite, the language L is regular.

Example 2.7. Check if the language L of nonempty binary words having the same number of a's and b's is regular. Namely, the language is defined as follows: $L = \{w \in \{a,b\}^* : \#_a w = \#_b w > 0\}$. For instance, $aabb \in L$, $babaab \in L$, $bab \notin L$.

Solution. The number of equivalence classes of the relation R_L induced by the language L is infinite. Consider the following infinite sequence of words:

$$\begin{aligned} w_1 &= a \\ w_2 &= aa \\ &\vdots \\ w_n &= a^n \\ &\vdots \end{aligned}$$

Notice that any two different words u and v of this sequence cannot belong to the same equivalence class: if $u = a^k$, $v = a^l$, $k \neq l$, then for $z = b^k$ we get $uz = a^k b^k \in L$ and $vz = a^l b^k \notin L$. Consequently, the relation R_L creates at least equivalence classes corresponding to words of the sequence, that is, infinitely many classes. In concluding, in light of the Myhill–Nerode lemma, the language L is not regular.

2.1.4 The pumping lemma

In addition to regular expressions and the Myhill–Nerode lemma, the pumping lemma is another tool that can be used to characterize regular languages. Regular expressions and the Myhill–Nerode lemma (and regular grammars, which are discussed in the next section) are aimed at proving that a language is regular. The Myhill–Nerode lemma and the pumping lemma can be used to prove that a language is not regular. As in the case of the Myhill–Nerode lemma, the pumping lemma is not proven here since important topics used in the proof have not been discussed yet. The lemma is formulated here since it is a crucial to the identification of regular languages. It will be proven in Chapter 8.

Lemma 2.2 (The pumping lemma for regular languages).
If *a language L is regular*
then *there exists a constant n_L such that for any word $z \in L$ the following condition holds:*

$$(|z| \geq n_L) \Rightarrow \left[\left(\bigvee_{u,v,w} z = uvw \wedge |uv| \leq n_L \wedge |v| \geq 1\right) \bigwedge_{i=0,1,2,\ldots} z_i = uv^iw \in L\right]$$

The pumping lemma formulates conditions necessary for a regular language. It shows that the nature of regular languages is finite and the length of words of a given regular language is limited by some constant n_L determined by the pumping lemma. If a regular language includes a word of length greater than or equal to n_L, then it is an infinite language. However, a structure of words that are longer than or equal to n_L is fairly simple. Such words are generated by inserting strings of a length limited by the constant n_L into words of the language that are shorter than n_L. That is, any word of a regular language not shorter than n_L has a floating part of being deleted, leaving the remaining part in the language.

Since the pumping lemma formulates necessary conditions, it is of limited practical usefulness in its direct form. It can be used for the analysis of the structure of words of the language. If words of a language satisfy the conclusion of the pumping lemma, then the language could be intuitively presumed to be regular. Then, based on such a supposition, the language could be proven to be regular. In practice, we use contrapositive of the pumping lemma rather than its generic version. The contrapositive of the pumping lemma formulates sufficient conditions for a language not to be regular. This makes contrapositive of the pumping lemma to help prove that specific languages are not regular.

Lemma 2.3 (Contrapositive of the pumping lemma for regular languages).
If *for any constant n_L there exists a word $z \in L$ such that*

$$(|z| \geq n_L) \wedge \left[\left(\bigwedge_{u,v,w} z = uvw \wedge |uv| \leq n_L \wedge |v| \geq 1\right) \bigvee_{i\in\{0,1,2,\ldots\}} z_i = uv^iw \notin L\right]$$

then *the language L is not regular.*

Example 2.8. Check that the language L defined in Example 2.7: $L = \{w \in \{a,b\}^* : \#_a w = \#_b w > 0\}$ is not regular one.

Solution. Let us employ the contrapositive of the pumping lemma. We assume a word $z = a^N b^N$ for a constant N not less than the constant n_L from the pumping lemma. For any split $z = uvw$ of the word z satisfying assumptions of contrapositive of the pumping lemma, the first two parts uv can include only a's: $u = a^p$, $v = a^q$, where $0 < q$ and $p + q \leq N$. Iterated words $z_i = a^p a^{(i-1)p} a^{N-(p+q)} b^N = a^{N-(i-1)q} b^N$ do not belong to the language L for any iteration value $i \neq 1$ (for instance, for $i = 0$ we get $z_0 = a^{N-q} b^N$). By contrapositive of the pumping lemma we conclude that the language L is not regular.

Example 2.9. Check if the language of binary words having number of zeros equal to square of a natural number is not regular:

$$L = \{w \in \{0,1\}^* : \#_0 w = n^2, n \in \{0,1,2,3,\ldots\}\}$$

Solution. Let us apply contrapositive of the pumping lemma. We assume a word $z = 0^{N^2} \in L$ for a constant $N \geq n_L$. For any split $z = uvw$ of the word z satisfying antecedent of contrapositive of the pumping lemma, that is, $0 < |v| \leq n_L \leq N$, let us take a particular iterated word $z_2 = 0^{N^2+|v|}$. The word $z_2 \notin L$ for the reason that $N^2 < N^2 + |v| \leq N^2 + N < (N+1)^2$. Again, by contrapositive of the pumping lemma, the language L is not regular.

2.1.5 Regular grammars

Regular grammars constitute the most straightforward class of grammars. They generate regular languages, so then they can be used for proving that languages are regular. In this section, we formulate a definition of regular grammars and use them as tools to generate regular languages. The formal proof that regular grammars generate regular languages will be given in Chapter 8.

Definition 2.4. A grammar $G = (V, T, P, S)$ is:
- left-linear if and only if all its productions take the form $A \to Bw$ or $A \to w$, where A, B are nonterminals, that is, $A, B \in V$, w is a string of terminals, possibly empty, that is, $w \in T^*$,
- right-linear if and only if all its productions of the form $a \to wB$ or $A \to w$, where A, B are nonterminals, w is a string of terminals, possibly empty,
- regular if and only if it is either left-linear, or right-linear.

Remark 2.5. Each left-linear grammar has its equivalent right-linear counterpart, and oppositely, each right-linear grammar has an equivalent left-linear counterpart. Equivalence of grammars is understood as the identity of generated languages. This observation will be proven in Chapter 8.

Example 2.10. The grammar given in Example 1.12 generates the language L of binary numbers without non-significant 0's. The grammar in this example is left linear, so then the language is regular. The following right linear grammar $G = (\{S, A\}, \{0, 1\}, P, S)$ generates the same language, where:

$$\begin{aligned} \text{P:} \quad & S \to 0 \mid 1A && (1), (2) \\ & A \to 0A \mid 1A \mid \varepsilon && (3), (4), (5) \end{aligned}$$

Solution. Indeed,

- words generated by the grammar G are included into the language L for the reason that they are:
  - either 0 (derived directly with the production (1) applied to the initial symbol S of the grammar, or
  - strings beginning with 1. Such strings are derived with the production (2) applied to the initial symbol S and then with any number of productions (3) and (4) applied to the nonterminal A. The derivation is terminated with the production (5),
- every word of the language L is generated in the grammar G. Notice that words of this language are either 0 or they begin with the digit 1. Hence:
  - 0 is derived as explained above: $S \to 0$,
  - any other word $1a_{k-1}a_{k-2}\ldots a_1a_0$ (a binary number with leading digit 1) can be derived applying the production (2) to the initial symbol S and then production (3) or (4) should be used for every next digit 0 or 1, respectively. The derivation is terminated with the production (5), when all letters have been generated. That is, a string $1a_{k-1}a_{k-2}a_{k-3}\ldots a_1a_0$ is generated as follow:

$$S \xrightarrow{1} 1A \xrightarrow{a_{k-1}+3} 1a_{k-1}A \xrightarrow{a_{k-2}+3} 1a_{k-1}a_{k-2}A \xrightarrow{a_{k-3}+3} 1a_{k-1}a_{k-2}a_{k-3}A \cdots$$
$$\xrightarrow{a_1+3} 1a_{k-1}a_{k-2}\ldots a_1 A \xrightarrow{a_0+3} 1a_{k-1}a_{k-2}a_{k-3}\ldots a_1a_0 A \xrightarrow{5} 1a_{k-1}a_{k-2}a_{k-3}\ldots a_1a_0$$

Example 2.11. Prove that the language of binary strings divisible by 5 is generated by a regular grammar. Note that binary strings permit non-significant 0's; for instance, 1, 01, 001 are valid binary strings.

Solution. Let assume that reminder of integer division of a binary number $a_ka_{k-1}\ldots a_1a_0$ by 5 is equal to r. Observe that the binary number $a_ka_{k-1}\ldots a_1a_0a$ (a digit a attached as least significant digit) divided by 5 gives remainder $(2r + a) \bmod 5$. Based on this observation we can build the following right-linear grammar generating natural numbers divisible by 5:

$$G = (\{S, A_0, A_1, A_2, A_2, A_3, A_4\}, \{0, 1\}, P, S)$$

$$\begin{array}{llll} \text{P:} & S \rightarrow 0A_0 \mid 1A_1 & & (1), (2) \\ & A_0 \rightarrow 0A_0 \mid 1A_1 \mid \varepsilon & & (3), (4), (5) \\ & A_1 \rightarrow 0A_2 \mid 1A_3 & & (6), (7) \\ & A_2 \rightarrow 0A_4 \mid 1A_0 & & (8), (9) \\ & A_3 \rightarrow 0A_1 \mid 1A_2 & & (10), (11) \\ & A_4 \rightarrow 0A_3 \mid 1A_4 & & (12), (13) \end{array}$$

The productions can be rewritten in shortened form:

$$\begin{array}{ll} \text{P:} & S \rightarrow 0A_0 \mid 1A_1 \\ & A_0 \rightarrow \varepsilon \\ & A_i \rightarrow 0A_{(2i) \bmod 5} \mid 1A_{(2i+1) \bmod 5} \quad \text{for } i = 0, 1, 2, 3, 4 \end{array}$$

For example, the derivation of the binary number 01010 is carried out as follows:

$$S \xrightarrow{1} 0\,A_0 \xrightarrow{4} 01\,A_1 \xrightarrow{6} 010\,A_2 \xrightarrow{9} 0101\,A_0 \xrightarrow{3} 01010\,A_0 \xrightarrow{5} 01010$$

This grammar generates the language L:

- information about the remainder of the division of the part derived so far is kept as the nonterminal's index. Derivation can be terminated with the production (5), which leaves a binary number divisible by 5. This confirms that any word generated in this grammar belongs to the language,
- simple inductive evidence shows that any word of the language can be generated in this grammar:
  - first of all, the grammar allows for generating all strings having one binary digit and supplemented by a nonterminal, that is, $0A_0$ and $1A_1$. It also allows to generate all strings of length one divisible by 5., that is, the only one string 0,
  - by the inductive hypothesis, let assume that the grammar allows for generating all binary strings of length n appended with respective nonterminal and allows for generating all binary strings of length n divisible by 5. Any binary string of length n appended to a nonterminal A_i can be extended to a binary string of length $n + 1$ by adding any digit (0 if the first A_i production is used and 1 for the second A_i production). This shows that any binary string of length $n + 1$ is generated. Then, the production (5) applied to the nonterminal A_0 generates a binary number of the length divisible by 5. Of course, all binary numbers of length $n + 1$ can be generated,
  - finally, due to the satisfaction of both inductive conditions, we can state that every word of the language is generated in the grammar.

Example 2.12. Design a left-linear grammar generating the language of binary strings divisible by 5, cf. Examples 2.10 and 2.11.

Solution. The right-linear grammar of Example 2.11 adds symbols at the end of a binary string. This grammar *controls* remainder of division by 5. A left-linear grammar adds symbols at the beginning of the string. Let us observe how the value of a binary number and the value of the remainder of integer division changes when a binary digit is added at the beginning of a binary string. Assume that a reminder of division of a binary number $a_k\, a_{k-1}\, a_{k-2} \dots a_1\, a_0$ by 5 is equal to r.

The digit 1 added at the beginning of a binary number $a_k\, a_{k-1}\, a_{k-2} \dots a_1\, a_0$ increases its value by 2^{k+1} and changes reminder of division by 5 to $(r + 2^{k+1}) \bmod 5$. Notice that $2^i \bmod 5$ is equal to 1, 2, 4, 3 when $i \bmod 4$ is equal to 0, 1, 2, 3, respectively. Notice also that $2^{i+1} \bmod 5 = (2 * (2^i \bmod 5)) \bmod 5$.

The digit 0 added at the beginning of the numbers neither changes its value nor remainder. However, it affects changes when the next possible digit is added.

Therefore, in order to be able to *control* remainder of division by 5, a left-linear grammar must *remember* position of added digit (more exactly – a reminder of division of this position's value by 5) and a reminder of division by 5 of the already generated string, and then apply conclusions of the above discussion. The discussion leads to the following grammar:

$$
\begin{aligned}
G = (\{S\} \cup \{&A_{i,j} : i,j \in \{0,1,2,3,4\}, i \neq 0\}, \{0,1\}, P, S) \\
\text{P:} \quad S &\to A_{2,0}\, 0 \mid A_{2,1}\, 1 \\
A_{i,0} &\to \varepsilon \\
A_{i,j} &\to A_{(2i) \bmod 5, j}\, 0 \mid A_{(2i) \bmod 5,\, (i+j) \bmod 5}\, 1 \\
&\quad \text{for } i,j \in \{0,1,2,3,4\}, i \neq 0
\end{aligned}
$$

There are 21 nonterminals: the initial symbol of the grammar S and 20 indexed nonterminals $A_{i,j}$. There are 46 productions: two are given explicitly, four $A_{i,0}$ are nullable productions and 40 are not nullable $A_{i,j}$ productions. The first index of a nonterminal $A_{i,j}$ stores remainder of division by 5 of the binary position of the nonterminal. The second index keeps remainder of division the remaining string of binary digits by 5. For instance, for $A_{i,j}\, a_k\, a_{k-1} \dots a_1\, a_0 : i = 2^{k+1} \bmod 5$ and $j = a_k a_{k-1} \dots a_1 a_0 \bmod 5$.

An example derivation of the binary number comes as follows:

$$
A \to A_{2,0}0 \to A_{4,2}10 \to A_{3,2}010 \to A_{10}1010 \to A_{2,0}01010 \to 01010
$$

The proof that this grammar generates the given language is similar to the proof we outlined in Example 2.11. It is left to the reader.

2.2 Problems

Problem 2.1. Prove equalities given in Remark 2.4.

Solution. 8. The equivalence $r(s + t) = rs + rt$ is understood in terms of equality of languages. Let assume that regular expressions r, s, t generate languages R, S, T, respectively. The regular expression $r(s + t)$ generates the language $R \circ (S \cup T) = \{uv : u \in$

$R \wedge v \in S \cup T\}$, which is equal to $R \circ (S \cup T) = \{uv : u \in R \wedge v \in S \vee u \in R \wedge v \in T\}$ and then it is equal to $\{uv : u \in R \wedge v \in S\} \cup \{uv : u \in R \wedge v \in T\} = R \circ S \cup R \circ T$. The latest language is generated by the regular expression $rs + rt$.

Problem 2.2. Prove that the language $L = \{w \in \{a,b\}^* : \#_a w > \#_b w > 0\}$ is not a regular one.

Solution. We apply the contrapositive of the pumping lemma to prove that L is not regular. Let N is a constant from the lemma and let $z = a^{N+1}b^N$. We consider any split $z = uvw$ satisfying antecedent of the lemma, that is, $|uv| \le N$ and $|v| \ge 1$. Thus, uv cannot include the letter b and v includes at least one letter a, that is, $u = a^p$ and $v = a^r$ where $1 \le r \le p + r \le N$. The word $z_0 = uv^0w = a^{N+1-r}b^N$ is not included in L. This observation completes the proof that the language is not regular.

Problem 2.3. Prove that the language $L = \{w \in \{a,b\}^* : \#_a w \neq \#_b w\}$ is not regular.

Solution. Let apply the Myhill–Nerode lemma. Consider the sequence of words: $u_1 = a, u_2 = aa, \dots, u_n = a^n, \dots$. Any two words of this sequence are not related in the relation R_L induced by L. Indeed, for any u_k and u_l, $k \neq l$, if – for instance – $z = b^k$, then $u_k z = a^k b^k \notin L$ and $u_l z = a^l b^k \in L$. This means that such two words belong to different equivalent classes of R_L and – as a result – we have an infinite sequence of equivalent classes of R_L.

Problem 2.4. Prove that the following language is regular

$$L = \{w \in \{a,b\}^* : \#_a w \bmod 3 = \#_b w \bmod 3\}.$$

Hint. Note: justify that there are three equivalence classes of the R_L relation: $E_i = \{w \in \{a,b\}^* : (\#_a w - \#_b w) \bmod 3 = i\}$, $i = 0, 1, 2$. Apply the Myhill–Nerode lemma.

Problem 2.5. Prove that the complement of a regular language $L \subset \Sigma^*$, that is, the language $\overline{L} = \Sigma^* - L$, is a regular language.

Hint. Notice that every equivalence class of the relation R_L induced by the language L is either included in L or disjoint with L and apply the Myhill–Nerode lemma.

3 Context-free grammars

The class of context-free languages is the next simplest class of languages, besides the class of regular languages. Context-free languages are generated by context-free grammars, which correspond to regular grammars and regular expressions. Context-free grammars are one of the few most important tools used for the analysis of context-free languages. They are incomparably simpler than context-sensitive and unrestricted grammars. On the other hand, they define a wider class of languages than simpler regular grammars. The structure of words they generate is rich in this sense that two parts of words can be simultaneously iterated, unlike in the case of regular languages, where one part could be iterated. Moreover, context-free grammars are well elaborated. They provide effective methods of analysis and generation of languages. Furthermore, there are effective methods of automatic analysis and processing of context-free grammars. Due to these advantages, they are widely applied in practice, e. g., in natural language processing, processing of programming languages, translation of formal languages, pattern recognition, etc.

Algebraic structures of context-free languages created in the set of all words over an alphabet are much more complex than those created by regular languages. For this reason, the algebraic analysis of context-free languages is limited. For instance, there is no tool corresponding to the Myhill–Nerode theorem.

This Chapter is devoted to a discussion on basic properties of context-free languages, especially those properties, which arise out of the analysis of context-free grammars. Let us emphasize that an analysis of context-free grammars is well established, given their practical relevance.

3.1 Context-free grammars – basics

Context-free grammars have a simple form of productions: a nonterminal creates production's left-hand side while a sequence of terminals and nonterminals figures its right-hand side. This form of productions allows for a simple illustration of derivation. A derivation of a word can be demonstrated as a tree, which easies the proofs of such important properties as the pumping lemma, the Ogden lemma and a decision algorithm for context-free languages.

Definition 3.1. A grammar $G = (V, T, P, S)$ is a context-free grammar if and only if left side of any its production is a nonterminal, that is, any $p \in P$ is of a form $A \to \alpha$, where $A \in V$ and $\alpha \in (V \cup T)^*$.

Notice that productions of regular grammars may have only one nonterminal in their right-hand side and those nonterminals must be at either left or right edge. So then, regular grammars are special case, straightforward case, of context-free grammars

https://doi.org/10.1515/9783112223123-004

Definition 3.2. A language $L \subset \Sigma^*$ is context-free if and only if it is generated by a context-free grammar.

Example 3.1. The following context-free grammar generates arithmetic expressions:

$$
\begin{aligned}
&G = (\{E\}, \{+.*, \mathrm{id}\}, P, E) \\
\text{P:}\quad & E \to E + E \qquad (1) \\
& E \to E * E \qquad (2) \\
& E \to \mathrm{id} \qquad\quad\; (3)
\end{aligned}
$$

The grammar generates a restricted form of arithmetic expressions with addition and multiplication only, not using brackets and with only a single argument id. Such grammar has limited practical importance. However, it will allow us for a convenient illustration of structures related to context-free grammars. On the other hand, it is easy to extend the grammar for subtraction and division (including new productions similar to the first two). Moreover, if the terminal id is replaced by a nonterminal and productions for this nonterminal are added, we get a more realistic grammar for arithmetic expressions.

Example 3.2. Let check that the word id $*$ id $+$ id is derivable from the initial symbol of the grammar provided in Example 3.1, that is, let build a derivation $E \xrightarrow{*} \mathrm{id} * \mathrm{id} + \mathrm{id}$. In the following derivation productions are applied to the underlined nonterminals:

$$\underline{E} \xrightarrow{1} \underline{E} + E \xrightarrow{2} \underline{E} * E + E \xrightarrow{3} \mathrm{id} * \underline{E} + E \xrightarrow{3} \mathrm{id} * \mathrm{id} + \underline{E} \xrightarrow{3} \mathrm{id} * \mathrm{id} + \mathrm{id}$$

where a number above the derivation symbol (right arrow) refers to an applied production, e. g. $\xrightarrow{1}$ informs that the first production is employed.

In this derivation, productions are always applied to the leftmost nonterminal of intermediate words of derivation. Such a derivation is called *leftmost derivation.*

A derivation with rightmost nonterminals yielding productions is called *rightmost derivation.* The following derivation of the word id $*$ id $+$ id is the rightmost one:

$$\underline{E} \xrightarrow{2} E * \underline{E} \xrightarrow{1} E * E + \underline{E} \xrightarrow{3} E * \underline{E} + \mathrm{id} \xrightarrow{3} \underline{E} * \mathrm{id} + \mathrm{id} \xrightarrow{3} \mathrm{id} * \mathrm{id} + \mathrm{id}$$

We can build another rightmost derivation of the word id $*$ id $+$ id:

$$\underline{E} \to E + \underline{E} \to \underline{E} + \mathrm{id} \to E * \underline{E} + \mathrm{id} \to \underline{E} * \mathrm{id} + \mathrm{id} \to \mathrm{id} * \mathrm{id} + \mathrm{id}$$

and yet another leftmost derivation of this word:

$$\underline{E} \to \underline{E} * E \to \mathrm{id} * \underline{E} \to \mathrm{id} * \underline{E} + E \to \mathrm{id} * \mathrm{id} + \underline{E} \to \mathrm{id} * \mathrm{id} + \mathrm{id}$$

In addition to the leftmost and rightmost derivations, there are several mixed derivations, that is, derivations in which productions are applied to non-systematically chosen nonterminals.

For context-free grammar, a word's derivation can also be presented in the form of a tree.

Definition 3.3. A derivation (parsing) tree in a context-free grammar $G = (V, T, P, S)$ is a tree $T = (W, E \subset W \times W)$ satisfying the following properties:
1. it is compatible with Definition 1.9, but the set of children of every vertex is ordered,
2. each vertex $w \in W$ of the tree is labeled with a nonterminal symbol or a terminal symbol or the empty word: $w \in V \cup T \cup \{\varepsilon\}$,
3. the root of the tree is labeled with the initial symbols S of the grammar,
4. internal vertices of the tree (that is, vertices different than leaves) are labeled with nonterminals,
5. if a nonterminal A labels an internal vertex and children of this vertex are labeled with symbols (terminals and nonterminals) $X_1 X_2 \dots X_k$ in this order or with the empty word ε, then there exists a production $A \to X_1 X_2 \dots X_k$ or the production $A \to \varepsilon$, respectively,
6. a vertex labeled with the empty word ε is the only child of its parent.

Notice that in this definition, we say *a vertex is labeled with a symbol* instead of *a vertex is a symbol* in order to handle cases of several occurrences of the same symbol in a tree.

We have two derivation trees for the word id $*$ id $+$ id, as shown in Figure 3.1. Both trees have the same crop, so they generate the same word. Notice that each derivation tree corresponds to one leftmost derivation and one rightmost derivation of this word.

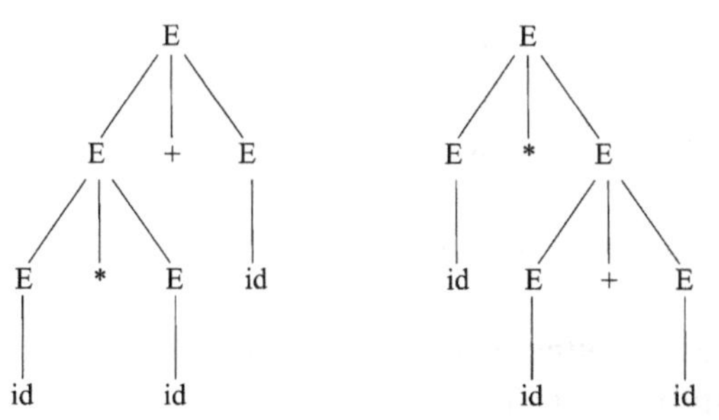

Figure 3.1: Derivation trees for the word id $*$ id $+$ id in Example 3.2.

Remark 3.1. Let us notice that the first derivation tree in Figure 3.1 corresponds to the first leftmost derivation and to the second rightmost derivation shown in Example 3.2. The second derivation tree of this figure corresponds to the second leftmost derivation and to the first rightmost derivation of the example. In general, every derivation tree has a corresponding leftmost derivation and rightmost derivation.

Derivation trees corresponding to derivations are independent of the order of nonterminals used for productions employed in a derivation. So then, there is no analogy between derivation trees and sided (leftmost/rightmost/mixed) derivations. However, if there is more than one derivation tree for a given word, then each derivation tree corresponds to a leftmost associated derivation and corresponds to an associated rightmost derivation. Therefore, in this meaning, there is correspondence between derivation trees, leftmost derivations and rightmost derivations.

We say that a word is ambiguous if it has more than a single derivation tree. Equivalently, we can say that a word is ambiguous if and only if it has more than one leftmost derivation or if it has more than one rightmost derivation. A context-free grammar is said to be ambiguous if and only if there is an ambiguous word. A context-free language is inherently ambiguous if and only if its every context-free grammar is ambiguous.

The grammar discussed in Example 3.1 is ambiguous. However, the language of arithmetic expressions is not inherently ambiguous. Later on in Part IV, we show an example of a nonambiguous context-free grammar for arithmetic expressions.

Example 3.3. The following language is inherently ambiguous:

$$\begin{aligned} L &= \{a^k b^l c^m : k, l, m, > 0 \wedge (k = l \vee l = m)\} \\ &= \{a^k b^k c^m : k, m, > 0\} \cup \{a^k b^m c^m : k, m, > 0\} \end{aligned}$$

Solution. A grammar $G = (V, T, P, S)$ for this language may have productions such as shown below:

$$\begin{array}{llll} \text{P:} & S \to L \mid R & & (1)(2) \\ & L \to Lc \mid Ac & & (3)(4) \\ & A \to aAb \mid ab & & (5)(6) \\ & R \to aR \mid aC & & (7)(8) \\ & C \to bCc \mid bc & & (9)(10) \end{array}$$

Any word of the form $w = a^k b^k c^k$ has two leftmost derivations (notice that productions have at most one nonterminal on the right-hand side):

$$S \xrightarrow{1} L \xrightarrow{(k-1)*3} Lc^{k-1} \xrightarrow{4} Ac^k \xrightarrow{(k-1)*5} a^{k-1}Ab^{k-1}c_k \xrightarrow{6} a^k b^k c^k$$

and

$$S \xrightarrow{2} R \xrightarrow{(k-1)*7} a^{k-1} R \xrightarrow{8} a^k C \xrightarrow{(k-1)*9} a^k Ab^{k-1} C c_{k-1} \xrightarrow{10} a^k b^k c^k$$

where formulas placed above derivation symbols (right arrows) identify a production employed and, in some cases, how many times it is yielded. For instance, $\stackrel{(k-1)*3}{\longrightarrow}$ means that production number 3 is applied $k-1$ times.

Words generated here are nonempty sequences of symbols a followed by symbols b and then followed by symbols c. There are two cases: lengths of the first two sequences (composed of symbols a and b) are equal or lengths of the last two sequences (composed of symbols b and c) are equal. Notice that the direct derivation from the initial symbol of the grammar (production number 1 or production number 2) decides which case is considered.

No formal proof of inherent ambiguity of this language is given here since this property is of minor importance for this book. The proof that some language is inherently ambiguous is given in [4]. Intuitively, every grammar for the language discussed here must have a mechanism deciding about the equality of lengths of the first two sequences or of the last two sequences. This mechanism raises ambiguity about such a grammar and – subsequently – of the language itself.

Example 3.4. Prove that the language $L = \{w \in \{a,b\}^* : \#_a w = \#_b w > 0\}$ is context-free.

Solution. The following context-free grammar generates L

$$G = (\{S\}, \{a,b\}, \{S \to ab \mid ba \mid aSb \mid bSa \mid SS \mid\}, S)$$

To show that the grammar G generates L, we will prove that any word of L has a derivation in G and then we will prove that any word generated in G is in L.

Before we attempt to do this, let as show that for any word $z \in L$, if it begins and ends with the same symbol, then it may be split to two parts, such that both belong to L. Formally, if $z_a = aw_a a$, $z_b = bw_b b$, $|w_a|, |w_b| > 0$ and $z_a, z_b \in L$, then there exist nonempty u_a, v_a, u_b, v_b such that $z_a = u_a v_a$, $z_b = u_b v_b$ and $u_a, v_a, u_b, v_b \in L$.

For a given word $z_a = a_1 a_2 \dots a_n$ (where $a_1 = a_n = a$ and $a_2, \dots, a_{n-1}$ are either a or b) define a mapping $\mu : \{1, 2, \dots, n\} \to N$ and for $k = 1, 2, \dots, n$, $\mu(k) = \#_a(a_1 a_2 \dots a_k) - \#_b(a_1 a_2 \dots a_k)$, that is, μ gives superiority of as over bs in a prefix of length k. Notice that $\mu(n) = 0$, what implies that $\mu(n-1) = -1$. Moreover, $\mu(1) = 1$ and $\mu(k)$ can increase or decrease its value by 1 for successive values of k. Consequently, μ must cross 0 somewhere in between 1 and $n-1$, that is, there must be $p \in \{2, 3, \dots, n-2\}$, for which $\mu(p) = 0$. As a result, $u_a = a_1 \dots a_p$ and $v_a = a_{p+1} \dots a_n$.

Consider a word z_b likewise.

Now, let us prove that any word $w \in L$ has a derivation in G:

- the shortest words of L are ab and ba. They are derivable with the first or the second production directly from the initial symbol S of G,
- for a longer word $z = a_1 w a_{n+2}$ consider its first and last symbols. There are four cases: (1) $a_1 = a$, $a_{n+2} = b$, (2) $a_1 = b$, $a_{n+2} = a$, (3) $a_1 = a = a_{n+2}$ and (4) $a_1 = b = a_{n+2}$.

The first and the second cases are similar and the third and the fourth cases are similar:
  - in the first and the second cases the sub-word w is in L. So, by the inductive assumption, there is a derivation $S \overset{*}{\to} w$ of w in G. So then $S \to aSb \overset{*}{\to} awb$ and $S \to bSa \overset{*}{\to} bwa$ are derivations of z in these cases,
  - in the third case $z = z_a = u_a v_a$, where $u_a, v_a \in L$, cf. the above discussion. Since $|u_a|, |v_a| < |z|$, then by inductive assumption, there exist derivations $S \overset{*}{\to} u_a$ and $S \overset{*}{\to} v_a$. Hence, $S \to SS \overset{*}{\to} u_a S \overset{*}{\to} u_a v_a$ is a derivation of z in G,
  - the fourth case is similar to the third one,
- based on mathematical induction we conclude that each word from the language L is generated in the grammar G.

Afterward, we prove that any word derivable in G belongs to L. Let $S \overset{n}{\to} z, z \in \{a, b\}^*$ is a derivation in G, where $\overset{n}{\to}$ denotes derivation of length n (productions are applied n times).
- for $n = 1$ a derivation produces either ab or ba, both in L,
- a derivation $A \overset{n+1}{\to} z, z \in \{a, b\}^*$ can be decomposed to:
  - either $A \to aSb \overset{n}{\to} awb$ or
  - $S \to bSa \overset{n}{\to} bwa$ or
  - $S \to SS \overset{k}{\to} uS \overset{l}{\to} uv$, where $1 \le k, l \le n$.

  By inductive assumption $w, u, v \in L$, that is, numbers of as and bs are equal in each of these words. Consequently, numbers of as and bs are equal in $z = awb$, $z = bwa$ and $z = uv$,
- based on mathematical induction we conclude that each word derivable in the grammar G is in the language L.

3.2 Simplification of context-free grammars

The definition of context-free grammars does not include optimization mechanisms, which would allow for simplification of grammars. In this section we discuss some methods of simplification of context free grammars. For instance, a context-free grammar including symbols and productions, which are never used in derivation of any word, may be transformed to a form simpler to use. Also, a context-free grammar can be restructured to a form more suitable for a given application.

3.2.1 Useless symbols

In particular, context-free grammar can include symbols that are *useless* for generated language. A set of nonterminals that will never produce a sequence of terminals is the first type of useless symbols, let us call them *not generative* ones. A set of symbols (terminals or nonterminals) never used in derivation from the initial symbol of the grammar forms the second type of useless symbols, call them *unreachable*. Both types of useless symbols can be removed together with productions including such symbols (such productions are invalid in terms of reduced sets of nonterminals and terminals). Both grammars, the initial one and the grammar with removed useless symbols and removed invalid productions, generate the same language. This observation is obvious since a production, which includes useless symbols, can never be used in the derivation of any word over the terminal alphabet.

Proposition 3.1. *For any context-free grammar $G = (V, T, P, S)$ generating a nonempty language $L(G)$ there exists an equivalent (that is, generating the same language) context-free grammar $G' = (V', T, P', S)$ such that every nonterminal $A \in V'$ generates a sequence of terminals (possibly empty). The following algorithm shows how to remove useless not generative symbols (symbols of the first type):*

```
begin
 V_old:=V_new:=∅
 for each production p: A→ α, p∈P do
   if α ∈T* then V_new:=V_new ∪{A}
     {start with nonterminals producing a string of terminals}
 while V_old ≠V_new do
 begin
   V_old:=V_new
   for each production p: A→ α, p∈P do
     if α ∈ (T∪V_old)* then V_new:=V_new ∪{A}
 end
 V':=V_new   {new set of nonterminals}
 P':={A→ α ∈ P : A ∈ V', supp(α) ⊂(V' ∪T)}
   {nonterminals of productions must be included into new set of nonterminals}
end
```

Notice that the initial symbol would be classified as useless if context-free grammar generates the empty language. However, any grammar must include an initial symbol, so then the initial symbol cannot be classified as a useless one.

Proposition 3.2. *For any context-free grammar $G = (V, T, P, S)$ generating a nonempty language $L(G)$ there exists an equivalent (that is, generating the same language) context-free grammar $G'' = (V'', T'', P'', S)$ such that every nonterminal symbol $A \in V''$ and every terminal symbol $a \in T''$ could be derived from the initial symbol of the grammar. The following algorithm allows for removing useless unreachable symbols (symbols of the second type):*

```
begin
 Vold:=Told:=∅
 Vnew:={S}, Tnew:=∅
 while Vold ≠Vnew or Told ≠Tnew do
 begin
  Vold:=Vnew, Told:=Tnew
  for A∈Vold do
  begin
    Vnew:=Vnew∪{B∈V : exists A→ α and B∈supp(α)}
    Tnew:=Tnew∪{a∈T : exists A→ α and a∈supp(α)}
  end
 end
 V'':=Vnew   {new set of nonterminals}
 T'':=Tnew   {new set of terminals}
 P'':={A→ α ∈ P : A ∈ V', supp(α) ⊂(V'' ∪T'')}
    {nonterminals and terminals of productions must be included
     into new sets of nonterminals and terminals}
end
```

Proof of Proposition 3.1 *and Proposition* 3.2. Let us observe that:

- a derivation tree of any word $w \in L(G')$ in the grammar G' is also a valid derivation tree in the grammar G. The same concerns the grammar G''. Therefore, $L(G') \subset L(G)$ and $L(G'') \subset L(G)$,
- a derivation tree of any word $w \in L(G)$ cannot have an internal node labeled with a useless not generative nonterminal (otherwise, there will be a leaf labeled with a nonterminal in a finite derivation tree). As a result, the derivation tree in G is also a valid derivation tree in G', that is, $L(G) \subset L(G')$,
- a derivation tree of any word $w \in L(G)$ cannot have a node (an internal node or a leaf) labeled with a useless unreachable symbol (otherwise, there will be no path from the root labeled with the initial symbol of the grammar G to this useless symbol). For that reason, the derivation tree in G is also a valid derivation tree in the grammar G'', that is, $L(G) \subset L(G'')$.

The above observation leads to the conclusion that $L(G) = L(G')$ and $L(G) = L(G'')$, what completes the proof. □

Removing useless symbols requires the application of the first algorithm followed by the second one, cf. [4]. The use of the algorithm in the opposite order may not remove all useless symbols, as shown in the following example.

Example 3.5. The following grammar:

$$G = (\{S, A, B\}, \{a\}, \{S \to SB \mid a, B \to AB, A \to a\}, S)$$

includes the not generative symbol B. Employing the first algorithm to remove useless symbols we get the grammar:

$$G'' = (\{S, A\}, \{a\}, \{S \to a, A \to a\}, S)$$

with a useless symbol A. The second algorithm produces the grammar with no useless symbols:

$$G'' = (\{S\}, \{a\}, \{S \to a\}, S)$$

The second algorithm employed prior to the first one does not change the grammar since it does not include any unreachable symbol of the second type. The first algorithm applied then produces the grammar

$$G = (\{S, A\}, \{a\}, \{S \to a, A \to a\}, S)$$

with a useless symbol A. The second algorithm employed again removes the remaining useless symbol.

Proposition 3.3. *The following property can be directly drawn from Proposition* 3.1 *and Proposition* 3.2*: any nonempty context-free language is generated by a context-free grammar without useless symbols.*

3.2.2 Nullable symbols and ε-productions

Elimination of ε-productions and nullable symbols is the next step of simplification of context-free grammars. In fact, we change the status of nullable symbols rather than eliminate them from grammars. Anyway, for the sake of simplicity, we will be using the term *eliminate*, but having this in mind that this means *change status* of nullable symbols.

ε-production is a production of the form $A \to \varepsilon$ and nullable symbol is a nonterminal symbol producing the empty word $A \xrightarrow{*} \varepsilon$. Of course, if the empty word is derivable from the initial symbol in a context-free grammar G, then it is not possible to eliminate all ε-production and nullable symbols. However, removing ε-productions and nullable symbols from the grammar G turns it to the grammar generating the language $L(G) - \{\varepsilon\}$. Therefore the process of elimination of nullable symbols and ε-productions will be applied to context-free grammars with the awareness that the empty word may be removed from the generated language.

The following algorithm finds nullable symbols in a context-free grammar $G = (V, T, P, S)$:

```
begin
 V_old :=∅   {begin with no nullable symbols}
 V_new :={A∈V : A→ ε is a production}
        {add all nonterminals producing directly the empty word}
 while V_old ≠V_new do
 begin
 V_old :=V_new
 V_new :=V_new∪{A∈V : exists A→ α, where α ∈V*_old}
     {add all nonterminals producing directly a word over nullable symbols}
 end
end
```

Having the set of nullable symbols, we will be able to remove ε-productions. Observe that nullable symbol in a production (its right-hand side) either can generate a string of terminal symbols or can be turned to the empty word. As a consequence, a nullable symbol can either be left on the right side of a production (when it produces a nonempty sequence of terminal symbols) or it can be dropped from the right-hand side of a production (when it generates the empty word). This observation leads to the following method.

Proposition 3.4. *Let $G = (V, T, P, S)$ is a context-free grammar with no useless symbols generating a language without the empty word. If $A \to X_1 X_2 \dots X_n$ is a production, then this production is replaced with a set of all productions of a form $A \to \alpha_1 \alpha_2 \dots \alpha_n$ that for all $i = 1, 2, \dots, n$ satisfies conditions:*
- *$\alpha_i = X_i$, if X_i is not nullable (that is, it is a terminal symbol or a not nullable nonterminal symbol),*
- *$\alpha_i = X_i$ or $\alpha_i = \varepsilon$, if X_i is a nullable symbol, that is, for a nullable symbol X_i we get two productions, one with X_i left at the right hand side and another one with X_i dropped from the right hand side,*
- *not all $\alpha_1, \alpha_2, \dots, \alpha_n$ are equal to the empty word (this condition eliminates ε-productions). Of course, the existing ε-productions are removed.*

Notice that this method turns status of nullable nonterminals to not nullable ones rather then removes them from the grammar. Finally, we get a grammar $G' = (V, T, P', S)$ without nullable symbols and ε-productions. This grammar has a modified set of productions and is equivalent to the grammar G, that is, generates the same language.

Proof. Proof of equivalence of both grammars is based on equivalence of derivations in both grammars G and G':
- both grammars have the same sets of terminals and nonterminals,
- a derivation in the grammar G of any word $w \in L(G)$ can be turned to a derivation of the same word in the grammar G':

- if a derivation does not employ ε-productions, then this is also a derivation in the grammar G',
- if an ε-production $X \rightarrow \varepsilon$ is applied in a part of derivation with a production $Y \rightarrow \alpha X \beta$ utilized prior to the $X \rightarrow \varepsilon$:

$$\ldots \rightarrow \gamma Y \delta \rightarrow \gamma \alpha X \beta \delta \rightarrow \gamma \alpha \beta \delta \rightarrow \ldots$$

  then this part can not be included in any derivation in G', but it can be turned to a fragment of a derivation in G' shown below. Here the production $Y \rightarrow \alpha\beta$ with the nullable symbol X dropped is employed,

$$\ldots \rightarrow \gamma Y \delta \rightarrow \gamma \alpha \beta \delta \rightarrow \ldots$$

  Finally, if we apply analogous replacement for every ε-production, the derivation of the word $w \in L(G)$ in the grammar G is turned into a derivation of the same word in the grammar G',

- a derivation in the grammar G' of any word $w \in L(G')$ can be turned to a derivation of the same word in the grammar G:
  - if a derivation utilizes only productions of the grammar G, then it is a valid derivation of the word w in G,
  - otherwise, a derivation employs at least one production of a form $Y \rightarrow \alpha\beta$ of the grammar G' gotten from a production of a form $Y \rightarrow \alpha' X \beta'$. Then a part of derivation

$$\ldots \rightarrow \gamma Y \delta \rightarrow \gamma \alpha \beta \delta \rightarrow \ldots$$

    can be replaced by a fragment of a derivation in a grammar G with inserted a series of ε-productions:

$$\ldots \rightarrow \gamma Y \delta \rightarrow \gamma \alpha' X \beta' \delta \rightarrow \cdots \rightarrow \gamma \alpha \beta \delta \rightarrow \ldots$$

    where $X \rightarrow \varepsilon$ is an ε-productions in the grammar G and strings α and β are gotten by applying the same scheme to all nullable symbols of both strings. This method applied to all productions of the derivation of the word $w \in L(G')$ in the grammar G' turns this derivation to a derivation of the same word in the grammar G.

Finally, comparing languages generated by a context-free grammar G and by its transformed form G' without nullable symbols and ε-productions, we can state that $L(G') = L(G) - \{\varepsilon\}$. □

Example 3.6. Remove nullable symbols and ε-productions from the grammar

$$G = (\{E, E_l, T, T_l, P\}, \{+, *, \mathrm{id}\}, P, E):$$

$$\begin{array}{llll} \text{P}': & E \to TE_l & & (1) \\ & E_l \to +TE_l & & (2) \\ & E_l \to \varepsilon & & (3) \\ & T \to PT_l & & (4) \\ & T_l \to *PT_l & & (5) \\ & T_l \to \varepsilon & & (6) \\ & P \to \text{id} & & (7) \end{array}$$

Solution. It is straightforward to show that the grammar does not include useless symbols and that the empty word cannot be derived. We have two nullable symbols: E_l and T_l. We take advantage of Proposition 3.4 to remove ε-productions and change the status of nullable symbols transforming productions of the grammar G to the following set of productions:

$$\begin{array}{lll} \text{P}: & E \to TE_l \mid T & (1) \\ & E_l \to +TE_l \mid +T & (2) \\ & \quad\text{not included} & (3) \\ & T \to PT_l \mid P & (4) \\ & T_l \to *PT_l \mid *P & (5) \\ & \quad\text{not included} & (6) \\ & P \to \text{id} & (7) \end{array}$$

3.2.3 Unit productions

A context-free grammar $G = (V, T, P, S)$ may have *unit productions*. Unit productions are of a form $A \to B$, where $A, B \in V$. Unit productions are confusing and do not provide any new abilities for language generation. Elimination of unit productions is the next step of grammar simplification. The method of removing unit productions is concerned with substituting a unit production $A \to B$ with a series of productions $A \to \alpha_i$ for all B-productions $B \to \alpha_i$. The method is outlined in the form of the following algorithm:

```
begin
 while there exists a unit production A→B do
   begin
    if (A=B) then remove the production
     else
       replace the production A→B with
       productions A→ γ1|...|γk
       where γ1,...,γk are right hand side
       of all B-productions
   end
end
```

The new grammar $G''' = (V, T, P''', S)$ produced by this algorithm may have useless nonterminal symbols, which need to be removed. For instance, compare [4], the grammar

$$G = (\{S, A, B\}, \{a\}, \{S \to A \mid a, A \to B, B \to a\}, S)$$

is turned to the grammar without unit production, but with useless nonterminal symbols A and B:

$$G''' = (\{S, A, B\}, \{a\}, \{S \to a, A \to a, B \to a\}, S)$$

Note that elimination of the unit production $S \to A$ introduces the production $S \to a$, which already exists in the grammar. The process of removing useless symbols yields the following grammar:

$$G^* = (\{S\}, \{a\}, \{S \to a\}, S)$$

A grammar without unit productions is equivalent to the former one. To prove this, let us consider a derivation tree in the former grammar. If a part of derivation tree matching a derivation $A_{i_1} \to A_{i_2} \to \cdots \to A_{i_r} \to \alpha$ with all productions except the last one are unit (the last production is not unit, the vertex A_{i_r} has at least two children or one leaf) includes a repeating nonterminal A' in the path $A_{i_1} \to \cdots A_{i_p} \to A' \to A' \to \cdots \to A' \to A' \to A_{i_r} \to \cdots A_{i_r}$, then this path can be shortened to $A_{i_1} \to \cdots A_{i_p} \to A' \to A_{i_r} \to \alpha$.

Let assume that a derivation $A_{i_1} \to A_{i_2} \to \cdots \to A_{i_r} \to \alpha$ with all productions except the last one are unit. The method of elimination of unit productions provides a production $A_{i_1} \to \alpha$, so then this derivation can be cut to $A_{i_1} \to \alpha$ which creates a part of a derivation tree in the grammar G'''. And vice versa, if we have a part of derivation tree in the grammar G''' matching a production $A_{i_1} \to \alpha$ in this grammar, then this production either belongs to the grammar G or it can be turned to a derivation $A_{i_1} \to A_{i_2} \to \cdots \to A_{i_r} \to \alpha$ in the grammar G.

In conclusion, a derivation tree in the grammar G can be turned to a derivation tree in the grammar G''' by replacing all such transformations. And vice versa, a derivation tree in the grammar G''' can be turned to a derivation tree in the grammar G.

Example 3.7. Remove unit productions from the grammar G' considered in Example 3.6.

Solution. There are two unit productions: the first one is in the group (1) of productions, the second one in the group (4) of productions. The first unit production $E \to T$ is replaced with two productions $E \to PT_l$ and $E \to P$. The production $E \to P$ is still unit one and is finally turned into $E \to \text{id}$. The second unit production $T \to P$ is replaced with the production $T \to \text{id}$. As a result we get the following set of productions:

$$
\begin{array}{llll}
\text{P:} & E \to TE_l \mid PT_l \mid \text{id} & & (1) \\
& E_l \to +TE_l \mid +T & & (2) \\
& \qquad \text{not included} & & (3) \\
& T \to PT_l \mid \text{id} & & (4) \\
& T_l \to *PT_l \mid *P & & (5) \\
& \qquad \text{not included} & & (6) \\
& P \to \text{id} & & (7)
\end{array}
$$

3.3 Normal forms of context-free grammars

We discuss two normal forms of context-free grammars, that is, Chomsky normal form and Greibach normal form. Conversions of context-free grammars to normal forms come as a further step in the simplification of grammars. Normal forms are grammars with restrictions imposed on the form of productions. Grammars in normal forms produce languages without the empty word, so then only grammars not generating the empty word could be transformed into normal forms. This is why a context-free grammar, in which the empty word is derivable, should be turned to a form generating the same language without the empty word. Since both normal forms do not admit ε-productions, removal of ε-productions and nullable symbols will convert any context-free grammar to a form generating the language without the empty word. Normal forms do not necessarily require the removal of useless symbols. Anyway, it is recommended to simplify a grammar by removing useless symbols first.

3.3.1 Chomsky normal form

Definition 3.4. A context-free grammar $G = (V, T, P, S)$ is in Chomsky normal form if and only if its productions are of the form $A \to BC$ or $A \to a$, where $A, B, C \in V$, $a \in T$ (that is, any production turns a nonterminal to two nonterminals or to one terminal).

Proposition 3.5. *Any context-free grammar without ε-productions and unit productions can be transformed to Chomsky normal form.*

Proof. Let assume that a context-free grammar $G = (V, T, P, S)$ does not have ε-productions and unit productions, so then a right-hand side of any production is a terminal symbol or is a string of at least two symbols. Productions with one terminal symbol on the right-hand side are in Chomsky normal form. Such productions will not be changed.

Productions having at least two symbols on the right-hand side will be transformed into a set of productions according to the following rules:

1. every production of a form $A \to \alpha_1\, a\, \alpha_2$ is substituted with two productions $A \to \alpha_1\, A'\, \alpha_2$ and $A' \to a$, where: $A \in V, a \in T, \alpha_1\alpha_2 \in (V \cup T)^+$ and A' is a new nonterminal. That is, each terminal (on the right hand side) is replaced with a new nonterminal and the production from this new nonterminal to replaced terminal is added,

2. every production of a form $A \to A_1 A_2 A_3 \dots A_n$, $n > 2$, $A_1, A_2, A_3, \dots A_n \in V$ is substituted with two productions $A \to A_{1,2} A_3 \dots A_n$ and $A_{1,2} \to A_1 A_2$, where $A, A_1, A_2, \dots, A_n \in V$, $A_{1,2} \in V$ is a new nonterminal. Notice that a production with right hand side of length $n > 2$ is replaced with a production in Chomsky normal form (two nonterminals) and a production with right hand side of length $n - 1$. Therefore, applying this operation $n - 2$ times to a production with right hand side of length n we turn it to $n - 1$ productions in Chomsky normal form.

The new grammar $G' = (V', T, P', S)$ includes newly added nonterminals. All productions of the grammar G not in Chomsky form are replaced with sets of new productions in Chomsky normal form.

Both grammars G and G' are equivalent, that is, they generate the same language. To show this, let us consider a derivation tree of a word in grammars G and G':

– a local fragment of a derivation tree in the grammar G matching a production of a form $A \to \alpha_1\, a\, \alpha_2$ is shown in part (i) of Figure 3.2. This fragment can be replaced with a fragment matching two productions $A \to \alpha_1 A' \alpha_2$ and $A' \to a$ (both substitute the former production) as shown in part (ii) of Figure 3.2. The new tree generates the same crop. On the other hand, a production $A \to \alpha_1 A' \alpha_2$ of the grammar G' forces the production $A' \to a$ since the nonterminal symbol A' is unique in the grammar G' and appears only in former two productions. Both later productions correspond to a part of a derivation tree in the grammar G' shown in part (ii) of Figure 3.2. This part can be turned to a structure shown in part (i) of Figure 3.2, which corresponds to a production $A \to \alpha_1\, a\, \alpha_2$ of the grammar G,

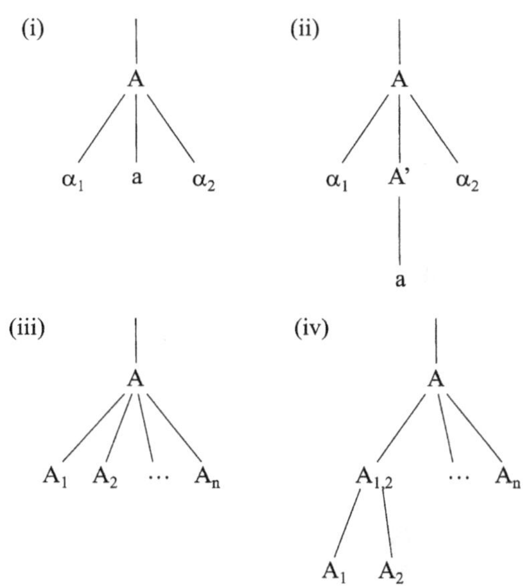

Figure 3.2: Equivalence of a context-free grammar and its Chomsky normal form.

- parts of derivation trees corresponding to a production of a form $A \to A_1 A_2 \dots A_n$ and to its substitutions $A \to A_{1,2} \dots A_n$ and $A_{1,2} \to A_1 A_2$ is shown in parts (iii) and (iv) of Figure 3.2,
- finally,
  - substituting all fragments of a derivation tree shown in parts (i) and (iii) of Figure 3.2 turns a derivation tree in the grammar G to a derivation tree in the grammar G',
  - opposite substitutions turns a derivation tree in the grammar G' to a derivation tree in the grammar G,
  - substitutions does not change the crop of subjected trees,

what justifies equivalence of grammars G and G'. □

Example 3.8. Find a grammar in Chomsky normal form for the set of production P'' considered in Example 3.7.

Solution.

- substitutions as in point 1 of Proposition 3.5 turns productions as shown below. Note that enumeration of groups of productions is inherited from the one shown in Example 3.6,

$$\begin{array}{rll} \mathrm{P}'': & E \to TE_l \mid PT_l \mid \mathrm{id} & (1) \\ & E_l \to \oplus TE_l \mid \oplus T & (2) \\ & \oplus \to + & (2') \\ & T \to PT_l \mid \mathrm{id} & (4) \\ & T_l \to \odot PT_l \mid \odot P & (5) \\ & \odot \to * & (5') \\ & P \to \mathrm{id} & (7) \end{array}$$

- substitutions as in point 2 of Proposition 3.5 turns productions to Chomsky normal form:

$$G_C = (\{E, E_l, T, T_l, P, T_\oplus, P_\odot, \oplus, \odot\}, \{+, *, \mathrm{id}\}, P_C, E)$$

$$\begin{array}{rll} P_C: & E \to TE_l \mid PT_l \mid \mathrm{id} & (1) \\ & E_l \to T_\oplus E_l \mid \oplus T & (2) \\ & T_\oplus \to \oplus T & (2'') \\ & \oplus \to + & (2') \\ & T \to PT_l \mid \mathrm{id} & (4) \\ & T_l \to P_\odot T_l \mid \odot P & (5) \\ & P_\odot \to \odot P & (5'') \\ & \odot \to * & (5') \\ & P \to \mathrm{id} & (7) \end{array}$$

Chomsky normal form of a context-free grammar is an important and powerful theoretical tool. The desired property of grammars in Chomsky normal form is that derivation trees are binary. Chomsky normal form will be exploited in the pumping lemma for context-free languages, in the Ogden lemma and in a decision algorithm for context-free languages.

3.3.2 Greibach normal form

Definition 3.5. A context-free grammar $G = (V, T, P, S)$ is in Greibach normal form if and only if its productions are of the form $A \to a\,\alpha$ where $A \in V$, $a \in T$, $\alpha \in V^*$ (that is, any production turns a nonterminal to a terminal and a string (possibly empty) of nonterminals.

Proposition 3.6. *Any context-free grammar without ε-productions and unit productions could be transformed to Greibach normal form.*

Proof. The following method can be used to transform a context-free grammar $G = (V, T, P, S)$ to Greibach normal form:
1. transform the grammar to Chomsky normal form $G_C = (V_C, T, P_C, S)$,
2. enumerate nonterminal symbols in $V_C = \{A_1, A_2, \ldots, A_n\}$,
3. modify the grammar G_C such that the right hand side of every A_i-production begins with a terminal symbol or with a nonterminal symbol of greater index:

$$(*) \quad \begin{cases} A_i \to a\alpha & \text{or} \\ A_i \to A_j\alpha \end{cases} \quad \text{where } a \in T, \quad A_i, A_j \in V_C \text{ and } j > i, \ \alpha \in V_C^*$$

Let assume that for $i = 1, 2, 3, \ldots, k-1$ all A_i productions satisfy the condition $(^*)$ given above and that some A_k production does not satisfy this condition. If a production $A_k \to A_j\,\alpha$, where $A_k, A_j \in V_C$ and $\alpha \in V_C^*$, does not satisfy $(^*)$, then either (i) $k > j$ or (ii) $k = j$. Then perform the following operations: (a) in the case (i) and (b) in the case (ii):

(a) for every A_j-production, $A_j \to \beta$, replace the production $A_k \to A_j\,\alpha$ with the production $A_k \to \beta\,\alpha$. The right-hand sides of new productions $A_k \to \beta_l\alpha$ either begin with a terminal or with a nonterminal A_l with index greater than j. Observe that every such substitution increases the value of the index of the right hand side nonterminal. Repeating substitutions at most $k - j$ times, we obtain productions replacing $A_k \to A_j\,\alpha$ with right-hand sides beginning with either a terminal symbol or with the nonterminal symbol A_k or with a nonterminal symbol A_l with $l > k$. Applying this operation to all A_k productions satisfying (i) we eliminate this case,

(b) let there is the following set of A_k-productions:

$$A_k \to A_k\,\alpha_1 \mid A_k\,\alpha_2 \mid \cdots \mid A_k\,\alpha_p \mid \beta_1 \mid \beta_2 \mid \cdots \mid \beta_r$$

where: $\alpha_1, \ldots, \alpha_p \in V_C^*, \beta_1, \ldots, \beta_r \in (T \cup \{A_{k+1}, \ldots A_n\}) \circ V_C^*$, that is, β_l is a terminal or a nonterminal with index greater than k followed by a sequence (possibly empty) of nonterminals.
This set of A_k-productions is replaced with the following set of new productions:

$$A_k \rightarrow \beta_1 \mid \beta_2 \mid \cdots \mid \beta_r \mid \beta_1 B_k \mid \beta_2 B_k \mid \cdots \mid \beta_r B_k$$
$$B_k \rightarrow \alpha_1 \mid \alpha_2 \mid \cdots \mid \alpha_p \mid \alpha_1 B_k \mid \alpha_2 B_k \mid \cdots \mid \alpha_p B_k$$

where: B_k, is a new nonterminal. Nonterminals B_i are ordered according to their indexes and are followed by all nonterminals A_i.
Notice that all newly included productions satisfy the condition (*). Therefore, the condition (*) is satisfied for all A_i-productions and B_i-productions for $i = 1, 2, \ldots, k$,

4. the process of the recent point repeated for successive nonterminals guarantee the satisfaction of the condition (*) for all productions. Moreover, the right-hand side of every A_n-production must begin with a terminal symbol since A_n is the last nonterminal in the introduced order, that is, every A_n-production is in the Greibach normal form,
5. for backward values of $k = n-1, n-2, \ldots, 2, 1$, for every A_k-production with right-hand side with a leading nonterminal symbol $A_j, j \in \{n, n-1, \ldots, k+1, \}$, replace this production with a new set of productions. Productions of this new set are obtained by replacing the leading nonterminal A_j with right hand sides of A_j-productions. We do the same with B_k-productions for decreasing values of index k. Productions of the newly created set are in Greibach normal form because all A_j-productions and B_j-productions have already been turned to Greibach normal form,
6. as a result, we get a grammar $G_G = (V_G, T, P_G, S)$ in Greibach normal form, where V_G is a set of nonterminal symbols V_C supplemented with nonterminal symbols B_k created in point 3(b).

It has already been noted that the conversion to Chomsky normal form does not change the language being generated. The transformation described in point 3(a) also keeps the generated language without any changes – justification is the same as for conversion to Chomsky normal form and for the elimination of unit productions.

We justify that elimination of looped productions in point 3(b) does not change the language. Let assume that a part of a derivation tree in the grammar G employs several productions of a form $A_i \rightarrow A_i \alpha_j$, that is, a nonterminal symbol A_i is substituted by a string of nonterminal symbols $A_i \alpha_i$ several times and finally A_i is substituted by a string β. The following example concerning a fragment of a derivation tree in the grammar G shown in part a) of Figure 3.3 is considered. This fragment is equivalent to the following derivation G:

$$A_i \rightarrow A_i \alpha_{i_3} \rightarrow A_i \alpha_{i_2} \alpha_{i_3} \rightarrow A_i \alpha_{i_1} \alpha_{i_2} \alpha_{i_3} \rightarrow \beta \alpha_{i_1} \alpha_{i_2} \alpha_{i_3}$$

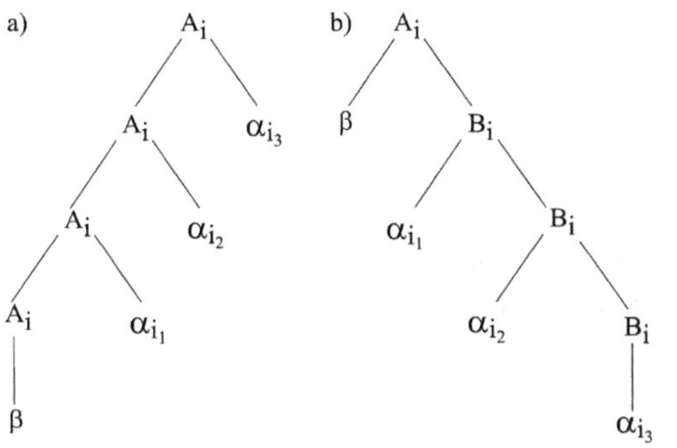

Figure 3.3: Invariability of unlooping method of a context-free grammar.

The rules in point 3(a) of Proposition 3.6 employed to the above fragment of derivation tree produces a fragment of a derivation tree in the grammar G_G. This fragment is shown in Figure 3.3(b). It is equivalent to the following derivation completed in the grammar G_G:

$$A_i \to \beta B_i \to \beta \alpha_{i_1} B_i \to \beta \alpha_{i_1} \alpha_{i_2} B_i \to \beta \alpha_{i_1} \alpha_{i_2} \alpha_{i_3}$$

Notice that other productions may distort derivations presented here. Nevertheless, the altered derivations are equivalent to the same derivation trees, cf. Remark 3.1. □

Example 3.9. Transform the given context-free grammar to Greibach normal form

$$G = (\{S, A, B\}, \{a, b\}, P, S)$$

$$\begin{aligned} \text{P:}\quad & S \to AB \\ & A \to BS \mid a \\ & B \to SA \mid b \end{aligned}$$

Solution. Assume the order S, A, B of nonterminals. Notice that all productions except $B \to SA$ satisfy the $(^*)$ condition of Proposition 3.6 Let us transform the grammar G to the Greibach normal form according to Proposition 3.6:

- the method 3(a) produces the following set of productions:

$$\begin{aligned} \text{P}':\quad & S \to AB \\ & A \to BS \mid a \\ & B \to ABA \mid b \end{aligned}$$

- the method 3(a) applied again produces the following set of productions:

$$\begin{aligned} \text{P}'':\quad & S \to AB \\ & A \to BS \mid a \\ & B \to BSBA \mid aBA \mid b \end{aligned}$$

- removing *looped* B-productions according to point 3(b) gives rise to the following set of productions with an extra nonterminal $\overline{B}$:

$$\begin{aligned} \text{P}''': \quad & S \to AB \\ & A \to BS \mid a \\ & B \to aBA \mid b \mid aBA\overline{B} \mid b\overline{B} \\ & \overline{B} \to SBA \mid SBA\overline{B} \end{aligned}$$

– now nonterminals are ordered as follows: $\overline{B}, S, A, B$. The backward process of point 5 brings Greibach normal form of the grammar G:

$$G_G = (\{S, A, B, \overline{B}\}, \{a, b\}, P_G, S)$$

$$\begin{aligned} P_G: \quad & B \to aBA \mid b \mid aBA\overline{B} \mid b\overline{B} \\ & A \to aBAS \mid bS \mid aBA\overline{B}S \mid b\overline{B}S \mid a \\ & S \to aBASB \mid bSB \mid aBA\overline{B}SB \mid b\overline{B}SB \mid aB \\ & \overline{B} \to aBASBBA \mid bSBBA \mid aBA\overline{B}SBBA \mid b\overline{B}SBBA \mid aBBA \mid \\ & \qquad aBASBBA\overline{B} \mid bSBBA\overline{B} \mid aBA\overline{B}SBBA\overline{B} \mid b\overline{B}SBBA\overline{B} \mid aBBA\overline{B} \end{aligned}$$

Example 3.10. Transform to Greibach normal form the given context-free grammar of Example 3.6:

$$G = (\{E, E_l, T, T_l, P\}, \{+, *, \text{id}\}, P, E)$$

$$\begin{aligned} \text{P}: \quad & E \to TE_l \mid T \\ & E_l \to +TE_l \mid +T \\ & T \to PT_l \mid P \\ & T_l \to *PT_l \mid *P \\ & P \to \text{id} \end{aligned}$$

Solution. Transformation of this grammar to Chomsky normal form will make the process more complex. So, we give up on the transformation of the grammar to the Chomsky normal form. Let order nonterminals as follows: E, E_l, T, T_l, P. Notice that all productions satisfy the $(^*)$ condition in Proposition 3.6. So then, employing the backward process of point 5 of Proposition 3.6 we get the following set of productions in the Greibach normal form:

$$\begin{aligned} \text{P}': \quad & E \to \text{id}\, TE_l \mid \text{id}\, E_l \mid \text{id}\, T_l \mid \text{id} \\ & E_l \to +TE_l \mid +T \\ & T \to \text{id}\, T_l \mid \text{id} \\ & T_l \to *PT_l \mid *P \\ & P \to \text{id} \end{aligned}$$

Finally, a grammar in Greibach normal form is called *simple*, if for any nonterminal symbol $A \in V$ and for any terminal symbol $a \in T$, there is at most one production $T \to a\,\alpha$, that is, there is at most one A-production with right-hand side beginning with given symbol a. This condition is called the Greibach uniqueness condition.

3.4 Pumping and Ogden lemmas

The pumping lemma for context-free languages and the Ogden lemma characterize the structure of words. Both lemmas are essential tools used in the identification of context-free languages.

3.4.1 The pumping lemma

The pumping lemma formulates conditions necessary for a language to be context-free. It shows that the nature of context-free languages is finite and the length of words of a given context-free language is limited by some constant n_L, whose value is determined in the pumping lemma. If a context-free language includes a word of length greater than or equal to n_L, then it is an infinite language. However, a structure of words that are longer than or equal to n_L, is fairly simple. Such words are generated by inserting strings of a length limited by the constant n_L into words of the language that are shorter than n_L. We can also say that any word of a context-free language, not shorter than n_L, have two floating parts that can be deleted simultaneously, leaving the remaining part in the language (let us recall that words of regular languages have only one part that can be subjected to deletion).

Lemma 3.1 (The pumping lemma for context-free languages).
If *a language L is context-free*
then *there exists a constant* n_L *such that for any word* $z \in L$ *the following condition holds:*

$$(|z| \geq n_L) \Rightarrow \left[\left(\bigvee_{u,v,w,x,y} z = uvwxy \wedge |vwx| \leq n_L \wedge |vx| \geq 1\right) \bigwedge_{i=0,1,2,\ldots} z_i = uv^iwx^iy \in L\right]$$

Proof. If a language L is finite, then a constant n_L greater than the length of the longest word of this language satisfies the lemma.

Consider the case of infinite languages. Let us assume that a context-free grammar $G = (V, T, P, S)$ in Chomsky normal form generates the language L. Derivation trees in such a grammar are binary trees. We use the property that the height (length of the longest path from the root to a leaf) of any binary tree with k leaves is not less than $\lceil \log_2 k \rceil$. All vertexes of such a path, except the last one, are labeled by nonterminal symbols of the grammar. The last vertex of this path is a leaf of the tree and is labeled by a terminal symbol.

Let $|V| = N$. If we set the constant $n_L = 2^N + 1$, then the height of a derivation tree of a word z not shorter than n_L is not less than $N+1$. Therefore there exists a path from the root S to a leaf not shorter than $N+1$. Consequently, $N+1$ vertexes of this path are labeled by N nonterminal symbols and the leaf is labeled by a terminal symbol a, cf. Figure 3.4. As a result, there exist two (maybe more) vertexes labeled by the same nonterminal symbol. Let us consider the pair of such vertexes closest to the leaf a that are labeled by the same nonterminal symbol A. To distinguish labels of vertexes of this pair, the nonterminal symbol A is denoted A' and A'' respectively. The crop z of the derivation tree is divided into five parts: u, v, w, x and y. The part w is the crop of the subtree with the root A'' while vwx is the crop of the subtree with the root A'.

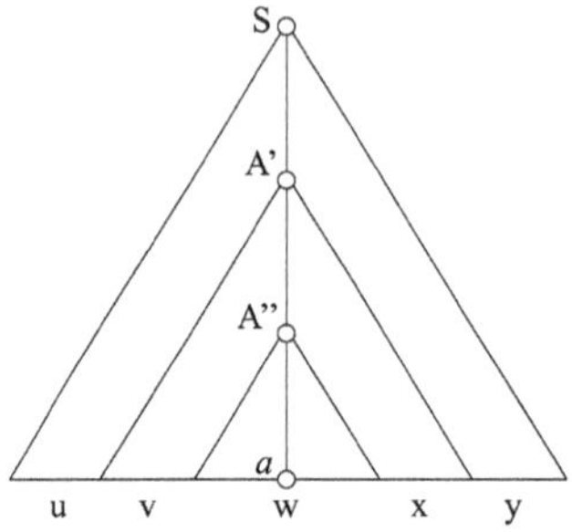

Figure 3.4: The derivation tree of a word z of length $2^{|V|+1}$.

If we replace the subtree with the root A' by the subtree with the root A'', we will get a valid derivation tree with the crop $z_0 = uwy = uv^0wx^0y$. This tree is shown at the left part of Figure 3.5. If we replace the subtree with the root A'' by the subtree with the root A', we will obtain the tree shown at the right part of Figure 3.5, which is still a valid derivation tree with the crop $z_2 = uvvwxxy = uv^2wx^2y$. Replacing the subtree with the root A'' in the last tree by the subtree with the root A', we get a derivation tree with the crop $z_3 = uvvvwxxxy = uv^3wx^3y$. This iterative process can be continued.

In this way, we show that the pumping lemma for context-free languages is satisfied. To fulfill formal requirements, mathematical induction should be applied based on the number of replacements of the subtree with the root A'' by the subtree with the root A'. Details of an inductive proof are left to the reader. □

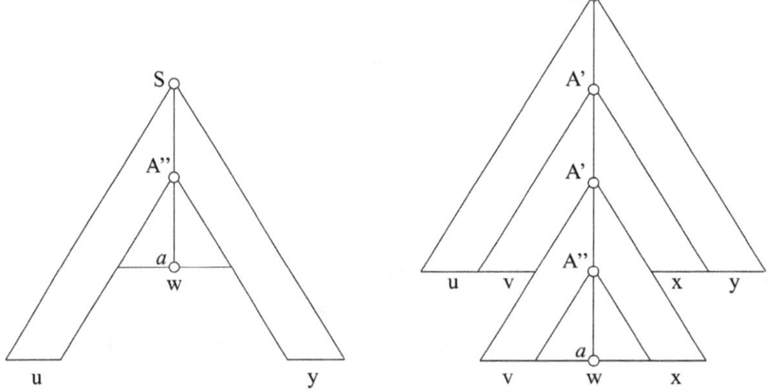

Figure 3.5: The derivation trees obtained from the tree shown in Figure 3.4.

Since the pumping lemma formulates necessary conditions, it is of limited practical importance in its direct form. It can be employed to analyze the structure of words of the language. If words satisfy the conclusion of the pumping lemma, then the language could be intuitively presumed to be context-free. Then, based on such a supposition, the language could be formally proved to be context-free. In practice, we use a contrapositive of the pumping lemma rather than its generic version, like in the case of the pumping lemma for regular languages. The contrapositive of the pumping lemma formulates sufficient conditions for a language not to be context-free. This makes the contrapositive to be very useful in proving that specific languages are not context-free.

Remark 3.2. Let us notice that the pumping lemma for regular languages is a special case of the pumping lemma for context-free languages. The assumption is that $uv = \varepsilon$ turns the pumping lemma for context-free languages to the pumping lemma for regular languages.

Lemma 3.2 (Contrapositive of the pumping lemma for context-free languages).
If *for any constant n_L there exists a word $z \in L$ such that*

$$(|z| \geq n_L) \wedge \left[\left(\bigwedge_{u,v,w,x,y} z = uvwxy \wedge |vwx| \leq n_L \wedge |vx| \geq 1 \right) \bigvee_{i \in \{0,1,2,\ldots\}} z_i = uv^i wx^i y \notin L \right]$$

then *the language L is not context-free.*

Example 3.11. Prove that the language $L = \{a^n b^{n+p} c^{n+p+q} : n, p, q \geq 1\}$ is not context-free.

Proof. We apply contrapositive of the pumping lemma. Let us take the word $z = a^N b^{N+1} c^{N+2}$, where N is the constant from the lemma. Note that vwx cannot include three different symbols, that is, a, b and c in any split of $z = uvwxy$ satisfying assumption of the lemma. Such a split satisfies one of the following conditions that are in pairs mutually exclusive:

- either v or x includes two different symbols,
- neither v, nor x includes the symbol a,
- neither v, nor x includes the symbol b or c. Therefore, they may include only as.

If a split satisfies the first condition, then $z_2 = uv^2wx^2y$ does not belong to the language L because this word has more than three sequences of symbols. In contrast, any word of the language L has exactly three sequences of symbols (namely, any word of the language has a sequence of symbols a followed by a sequence of symbols b followed by a sequence of symbols c).

If a split satisfies the second condition, then iterated parts v and x include either bs, or cs, or both symbols. Taking $z_0 = uv^0wx^0y = uwy$ we decrease either the number of bs, or the number of cs, or the numbers of both symbols. Of course, the number of as is unchanged. In the first and the third cases, there are no more bs than as in z_0. In the second case, there is no more cs than bs in z_0. Therefore, $z_0 \notin L$.

For the third condition, iterated parts v and x include as, or bs, or both symbols. Taking $z_2 = uv^2wx^2y = uwy$, we increase the number of as. Of course, the numbers of bs and cs are unchanged. In consequence, there is no more bs than as in z_0. Finally, $z_2 \notin L$.

Finally, since premises of contrapositive of the pumping lemma are satisfied, we conclude that the language is not context-free. □

3.4.2 The Ogden lemma

The pumping lemma, its contrapositive, is a powerful tool used to prove that languages are not context-free. However, the contrapositive of the pumping lemma can hardly be applied for some types of languages. In such complex cases, the Ogden lemma may help in proving that a language is not context-free. The pumping lemma is a particular case of the Ogden lemma. Nevertheless, the pumping lemma is easier to be applied than the Ogden lemma. This is why the pumping lemma is used for more straightforward problems.

Lemma 3.3 (The Ogden lemma).
If *a language L is context-free*
then *there exists a constant n_L such that for any word $z \in L$ and for at least n_L symbols marked in z there exists a split $z = uvwxy$ satisfying the conditions:*
– *vx includes at least one marked symbol,*
– *vwx includes no more than n_L marked symbols*
and such that $z_i = uv^iwx^iy \in L$ for any $i = 0, 1, 2, \ldots$.

Proof. Let us notice that the height of a derivation tree is, of course, not less than $\lceil \log_2 n_L \rceil$. This means that there exists a path from the root to a leaf not shorter than $\lceil \log_2 n_L \rceil$. Such a path starts in the root and then, for each its vertex, goes to this child, which has no less marked leaves in its subtree than another one has. Having such a path, we can do the same replacements in the derivation tree as we did in the proof of the pumping lemma. □

Notice that by marking all symbols of a word, we turn the Ogden lemma into the pumping lemma. Thus, the pumping lemma is a special case of the Ogden lemma.

As in the case of the pumping lemma, the Ogden lemma formulates necessary conditions for a language to be context-free. This is why the Ogden lemma in its direct form is hardly applicable, like pumping lemmas. In practice, we use the contrapositive of the Ogden lemma as well.

Lemma 3.4 (Contrapositive of the Ogden lemma).

If *for any constant n_L there exists a word $z \in L$ with at least n_L symbols marked symbols marked such that for any split $z = uvwxy$ holding conditions:*
- *vx includes at least one marked symbol,*
- *vwx includes no more than n_L marked symbols*

and such that there exists a constant $i \in \{0, 1, 2, \ldots\}$ for which $z_i = uv^iwx^iy \notin L$

then *the language is not context-free.*

Example 3.12. Prove that the following language is not context-free, cf. [4]:

$$L = \{a^k b^l c^m : k, l, m > 1 \text{ and } m \neq kr \wedge m \neq ls \text{ where } r, s \in \{2, 3, 4, \ldots\}\}.$$

Solution. The contrapositive of the pumping lemma is not applicable to this language (or, at least, we do not know how to employ it). On the other hand, trials to find a context-free grammar generating this language fail. This suggests (but does not prove) that the language might not be context-free, despite that the pumping lemma is not applicable for it. So then, let us apply the contrapositive of the Ogden lemma to prove that this language is not context-free. The idea of this proof relies on iterating a sequence of symbols c in order to get their number to be a multiple of the number of symbols a or the number of symbols b. We can do it since the Ogden lemma allows to force symbols c to be included in iterated parts vx of the split $uvwxy$, while in the pumping lemma such a forcing is impossible.

Assuming that n_L is the constant form contrapositive of the Ogden lemma, let us take a prime number N such that it is greater than n_L and greater than 3. In the word $z = a^N b^N c^{(N-1)!}$ all symbols c are marked. Since N is prime, then it is not a divisor of $(N-1)!$. This is why $z \in L$.

Let consider splits of the word $z = uvwxy$. Any iterated part (v or x) can include (multiples of) three different symbols or (multiplies of) two different symbols or a (multiple of) one symbol or is empty (though one of v and x must be a nonempty word). We can have the following types of splits:

1. an iterated part (v or x) includes three or two different symbols, that is, a and b and c or a and c or b and c. Notice that only symbols c are marked, then they must appear in at least one repeated part,
2. iterated parts include only symbols c,
3. the iterated part v contains only symbols a and the iterated part x contains only symbols c. Notice that the case that two different symbols appear in a repeated part is considered in the first point. Notice also that since symbols c must appear, they must be in x,
4. the iterated part v contains only symbols b and the iterated part x contains only symbols c.

For any type of splits there exists an iteration coefficient i, which takes the word $z_i = uv^iwx^iy$ out of the language L. Following iterations take the iterated word out of the language L for splits listed above:

1. the word $z_2 = uv^2wx^2y \notin L$ because symbol b precedes symbols a or symbol b precede symbols c,
2. let observe that iterated parts vx includes q symbol c, where $1 \le q \le n_L$. We look for such a coefficient i that $z_i = uv^iwx^iy \notin L$. Notice that $z_i = a^N b^N c^{(N-1)!+(i-1)q}$, what means that we wish N dividing $(N-1)! + (i-1)q$, that is, we look for such a factor r that $rN = (N-1)! + (i-1)q$. A simple conversion gives the value of the iteration coefficient equal to:

$$i = \frac{rN - (N-1)!}{q} + 1$$

   and for $r = (N-1)!$ we get

$$i = \frac{(N-1)(N-1)!}{q} + 1$$

   Since $1 \le q \le n_L < N$, then q divides $(N-1)!$, so then the last fraction is a positive integer number, which provides an iteration coefficient i excluding z_i from the language L,
3. iterations do not change the number of symbols b. So then, the value of the iteration coefficient, as computed in case 2, gives the number of symbols c to be the multiple of the number of symbols b,
4. the same as in the case 3, but with regard to symbols c and a.

Finally, based on the contrapositive of the Ogden lemma, since for any split we can take the iterated word out of the language, this language is not context-free.

3.5 Context-free language membership

The central question is how to check if a word belongs to a language. Having context-free grammar, we can answer the question if a word is generated in this grammar. Moreover, we will be able to build a derivation of a given the word if it is generated in the grammar.

First of all, let assume that grammar is in Greibach normal form. Note that any production applied to a derivation adds a single terminal symbol. This means that any derivation of a word of length n has length n (that is, productions are applied n-times in a derivation). A method of building a derivation is simple. Of course, we start with the initial symbols S of the grammar and apply a S-production with the right-hand side beginning with the first symbol of the word. In the next steps, we take the leftmost non-terminal symbol A of an intermediate derivation word and the next consecutive symbol a of the word and apply an A-production beginning with the terminal symbol a.

If a grammar satisfies Greibach uniqueness condition, that is, for every nonterminal symbol A and for every terminal symbol a, there is at most one A-production with the right-hand side beginning with a; then there is no ambiguity in the choice of productions. Otherwise, when the grammar does not satisfy the uniqueness condition, there is a question of how to choose a production. We either can make a nondeterministic choice between all A-production with the right-hand side starting with a given terminal symbol, or can check if any possible derivation produces the given the word w. In case of checking all possible derivations, assuming that we have no more than k productions for the choice, we may have up to k^n derivations, where the length of the word w is equal to n. This means that the computational complexity of this method is exponential, which makes it useless in practice. Note: a concept of nondeterminism will be discussed in further parts of the book.

There are more algorithms for membership tests, many of them being applicable to special forms of context-free grammars. We discuss here an algorithm invented by J. Cocke, H. Younger and T. Kasami. The algorithm is called the Cocke–Younger–Kasami algorithm or the CYK algorithm for short. The CYK algorithm operates on context-free grammar in Chomsky normal form.

The way of determining whether a word w of length n is generated in a grammar G in Chomsky normal form is outlined as follows:

1. split the word w into n substrings of length one (each symbol of the word w makes up a substring of length one, in this case) and find out nonterminal symbols generating each substring. This operation is a simple lookup for productions of the form $A \to a$, where a is a given substring of length one,
2. having nonterminal symbols generating substrings of the word w not longer than k, we can find nonterminals generating substrings of length $k+1$ as follows
   a. split a substring of length $k+1$ to all possible pairs of substrings (that is, prefixes of lengths $1, 2, 3, \dots, k$ and the corresponding suffixes of length $k, k-1, k-3, \dots, 2, 1$),
   b. for every pair of a prefix and the corresponding suffix sets of nonterminal symbols generating them have already been determined,
   c. find out the set of all nonterminals A such that there is a production $A \to BC$, where B and C are nonterminals generating the prefix and the corresponding suffix,
   d. nonterminals found out in the point c generate the given string of length $k+1$ and no other nonterminal does,
3. finally, we get the set of all nonterminal symbols generating the substring of length n, that is, generating the word w. The word w is generated in the grammar if and only if the initial symbol of the grammar is in this set.

Assuming that $w = a_1 a_2 \dots a_n$ is an analyzed word and V_i^j is the set of all nonterminals generating the substring $a_i a_{i+1} \dots a_{i+j-1}$ of the word w, for $j = 1, 2, \dots, n$, $i = 1, 2, \dots, n-j+1$, the Cocke–Younger–Kasami algorithm can be formulated as follows:

```
begin
1. find out all sets V_i^1 of nonterminal symbols
         generating the symbol a_i of the word w
2. for consecutive length values j = 2,3,...,n
            of substrings of the word w do
3.   for k = 1,2,...,n − (j − 1) do
     begin
4.     initialize sets V_k^j to the empty set
           and take the substring a_k ... a_{k+(j−1)}
5.     for splits of a_k ... a_{k+(j−1)} to prefix a_k ... a_{k+(l−1)}
             and suffix a_{k+l} ... a_{k+(j−1)}, l = 1,2...j − 1 do
      begin
6.      find out all productions A → BC
            s.t. B ∈ V_k^l and C ∈ V_{k+l}^{j−l}
7.      include all such A's into the set V_k^j
      end
     end
8. the word w is generated in the grammar
      if and only if the initial symbol of the grammar
      is included into the set V_1^n
```

Notice: for easier analysis of the above algorithm let us comment meaning of:

– V_k^j is the set of all nonterminals generating the sub-string of length j:

$$a_k \ a_{k+1} \ a_{k+2} \ \dots \ a_{k+(l-1)} \ a_{k+l} \ a_{k+(l+1)} \ \dots \ a_{k+(j-1)}$$

– V_k^l is the set of all nonterminals generating prefix of length l of the above sub-string:

$$a_k \ a_{k+1} \ a_{k+2} \ \dots \ a_{k+(l-1)}$$

– V_{k+l}^{j-l} is the set of all nonterminals generating suffix of length $j - l$ of the above sub-string:

$$a_{k+l} \ a_{k+(l+1)} \ a_{k+(l+2)} \ \dots \ a_{k+(j-1)}$$

Example 3.13. Verify if the word $w = abcab$ is generated in the grammar $G = (\{S, A, B, C, D\}, \{a, b, c, d\}, P, S)$ with productions:

$$\begin{aligned} \text{P:}\quad & S \to AB \mid CD \mid DB \\ & A \to BC \mid a \\ & B \to CD \mid b \\ & C \to AA \mid DC \mid c \\ & D \to AB \mid d \end{aligned}$$

Solution. The result of the CYK algorithm is shown in Table 3.1. Entries of the table are filled in with sets of nonterminal symbols V_k^j. The word is generated in the grammar since the set V_1^5 (bottom right entry of the table) includes the initial symbols of the grammar.

Table 3.1: Results of computation of the CYK algorithm for the word *abcab* in Example 3.13.

	a	*b*	*c*	*a*	*b*
V_k^1	A	B	C	A	B
V_k^2	S, D	A	$\emptyset$	S, D	
V_k^3	C	C	S, B		
V_k^4	$\emptyset$	$\emptyset$			
V_k^5	S, B				

Let us analyze the computational complexity of the CYK algorithm. Observe that costs of the following operations are upper-bounded by constants:

- initialization of sets V_i^1 in operation 1,
- getting a substring in operation 3,
- initialization of sets V_k^j to empty sets in operation 4,
- splitting a string into a prefix and a suffix in operation 5,
- finding out productions in operation 6 (which requires checking all production of the grammar and since the number of production is fixed, then cost of this operation is bounded by a constant),
- including left hand sides A of productions into sets in operation 7.

Finding out productions in operation 6 is a dominant operation of this algorithm. The number of executions of this operation is equal to:

$$\begin{aligned}\sum_{j=2}^{n}\sum_{k=1}^{n-j+1}(j-1) &= \sum_{j=2}^{n}(n-j+1)(j-1)\\ &= -\sum_{j=2}^{n}j^2 + (n+2)\sum_{j=2}^{n}j - (n-1)(n+1)\\ &= -\left(\frac{n^3}{3}+\frac{n^2}{2}+\frac{n}{6}-1\right) + \frac{(n+2)(n+2)(n-1)}{2} - n^2 + 1 = \frac{n^3-n}{6}\end{aligned}$$

what means that the complexity of the CYK algorithm is of the range $\Theta(n^3)$ regarding the length of an analyzed word.

The basic version of the CYK algorithm is used to find out if a word is generated in the grammar, but it does not allow for finding a derivation tree. A modified version of the CYK algorithm, with the extended version of operations 6 and 7, gathers information necessary for building derivation trees:

```
6.  find out all productions A → BC
       s.t. B ∈ V_k^l and C ∈ V_{k+l}^{j-l}
7.  include all such A's into the set V_k^j
7a. store right hand side BC of the production A → BC
       and parameter l as the subscript A_→ attached
       to the left hand side A of the production
```

The above operation 7a attaches the right-hand side of every A-production determined by operation 6 to the nonterminal symbol A. It means that every nonterminal symbol A in every set V_k^j has some associated set $A_{\rightarrow}$ of the right-hand side of an A-production generating the corresponding substring of the word w. The following algorithm generates a derivation tree based on the results of the extended CYK algorithm. Parameters of the function `generate` (generate a subtree of the derivation tree) denote `k` – the position in the word of the first symbol of the substring, `j` – length of the substring, `A` – the nonterminal generating the substring, `here` – the position of the nonterminal `A` in the tree.

```
generate(k,j,A,here)
begin
 if j=1 then
 begin
   put the symbol A at the place here,
   draw an edge from A down to the terminal symbol
 end
 else
 begin
   newTree:=false
      /*currently built tree is the current copy*/
   for every BC, l ∈ A_→ do
   begin
     if newTree then
     begin
       create a copy of the derivation tree
       built before the current call
       of this function, make newly created copy
       to be the current copy,
     end
     apply subsequent operations to the current copy
     put the symbol A at the place here
     draw a left edge to a leftVertex
```

```
      call generate(k,1,B, leftVertex)
      draw a right edge to a rightVertex
      call generate(k+1,j-1,C, rightVertex)
      newTree := true
    end
  end
end
```

The first call of the function `generate` requests building the whole derivation tree for an investigated word w of length n and is as follows:

```
generate(1,n,S,position-of-the-root).
```

Example 3.14. Find derivation trees for Example 3.13.

Solution. Results of computation of the extended CYK algorithms are displayed in Table 3.2. Sets $A_{\rightarrow}$ are displayed as subscripts of nonterminals symbols of sets V_k^j. The word w has three derivation trees shown in Figure 3.6, that is, derivation of this word is ambiguous.

Table 3.2: Results of computation of the extended CYK algorithm for the word *abcab* in Example 3.13.

	a	*b*	*c*	*a*	*b*
V_k^1	$A_{\{a\}}$	$B_{\{b\}}$	$C_{\{c\}}$	$A_{\{a\}}$	$B_{\{b\}}$
V_k^2	$S_{\{AB,1\}}, D_{\{BC,1\}}$	$A_{\{BC,1\}}$	$\emptyset$	$S_{\{AB,1\}}, D_{\{AB,1\}}$	
V_k^3	$C_{\{AA,1,DC,2\}}$	$C_{\{AA,2\}}$	$S_{\{CD,1\}}, B_{\{CD,1\}}$		
V_k^4	$\emptyset$	$\emptyset$			
V_k^5	$S_{\{CD,3,DB,2\}}, B_{\{CD,3\}}$				

The first and the second derivation trees are built on the basis of the production $S \rightarrow CD$ applied to the initial symbol of the grammar, which is the root of the derivation tree. The third derivation tree is built on the production $S \rightarrow DB$ applied to the initial symbols of the grammar. The difference between the first and the second derivation trees stems from the ambiguity of the left nonterminal C derived from the initial symbol of the grammar, cf. left subtrees of the first and the second trees.

3.6 Problems

Problem 3.1. Design context-free grammars generating the following languages:
1. the set of palindromes over the alphabet $\{a, b\}$,

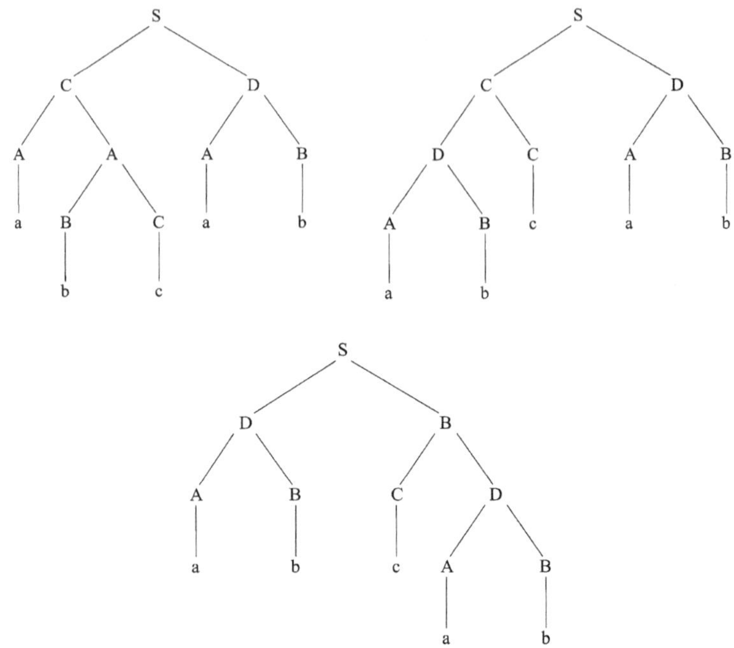

Figure 3.6: The derivation trees produced by the extended CYK algorithm.

2. the set of correct sequences of parentheses,
3. the set of nested brackets (brackets, square brackets, round brackets),
4. the set of nonempty binary words having the same numbers of as and bs,
5. the set of binary words having twice as much as than bs,
6. the set of regular expressions over the alphabet $\{a, b\}$.

Solution. 4. (Compare also solution of Example 3.4.) The following formula $L = \{w \in \{a, b\}^* : \#_a w = \#_b w > 0\}$ defines the language. The following inquiry leads to a solution. The initial symbol S of the grammar generates words having the same numbers of as and bs, that is, $S \xrightarrow{*} w$. In a word $w = a\, a_2 \dots a_n \in L$, which begins with a, the string $a_2 \dots a_n$ has the number of bs exceeding the number of as by 1. Assume that strings with the number of bs exceed the number of as by 1 are generated by the nonterminal B. By analogy, consider the case $w = ba_2 \dots a_n \in L$ and assume that strings with the number of as exceeds the number of bs by 1 are generated by the nonterminal A. So, the following productions $S \to aB \mid bA$ could be employed.

Now, consider strings $u = a\, a_2\, a_3 \dots a_n$ generated by the nonterminal A, that is, $A \xrightarrow{*} u$. In this case the string $a_2\, a_3 \dots a_n$ has the same number of as and bs. On the other hand, in a string $u = ba_2\, a_3 \dots a_n$ generated by the nonterminal A, the number of as exceeds the number of bs by 2. These cases employs productions $A \to a \mid aS \mid bAA$. Likewise, productions $B \to b \mid bS \mid aBB$ should be considered.

Finally, the above intuitive investigation brings the following grammar $G = (\{S, A, B\}, \{a, b\}, P, S)$ with the following set P of productions:

$$\begin{aligned} \text{P:}\quad & S \to aB \mid bA \\ & A \to a \mid aS \mid bAA \\ & B \to b \mid bS \mid aBB \end{aligned}$$

A proof that this grammar generates the language L can use mathematical induction concerning words with the same number of as and bs, with the number of as exceeding the number of bs by 1 and with the number of bs exceeding the number of as by 1. We claim that all such words and only such words are generated by nonterminals S, A, B, respectively. Notice that words with the same number of as and bs have even length, that is, length equal to $2n$ for $n = 1, 2, 3, \ldots$. Similarly, length of words of other two types is odd, that is, is equal to $2n - 1$ for $n = 1, 2, 3, \ldots$.

For $n = 1$ we have the following words of the above three types: a, b, ab, ba. These words are derivable from corresponding nonterminals as follows: $A \to a$, $B \to b$, $S \to aB \to ab$ and $S \to bA \to ba$. Of course, no other word of the assumed lengths could be derived.

Based on inductive hypothesis we assume that for a given n

- for all $k = 1, 2, \ldots, n$ all words of length $2k - 1$ having k symbols a and $k - 1$ symbols b, and only such words, are derivable from A,
- for all $k = 1, 2, \ldots, n$ all words of length $2k - 1$ having $k - 1$ symbols a and k symbols b, and only such words, are derivable from B,
- all words of length $2k$ having k symbols a and k symbols b, and only such words, are derivable from S.

Doing inductive step we get:

- any word of length $2n+1$ having $n+1$ symbols a and n symbols b is derivable from A. Such a word
  - either begins with a and has n (equal numbers) of as and bs in the remaining part. So, based on the inductive assumption, we get that the production $A \to aS$ allows to generate it,
  - or it begins with b and has $n + 1$ symbols a and $n - 1$ symbols b in the remaining part. The remaining part can be split into two parts, in which the number of as exceeds the number of bs by 1 (the reader can employ the function μ used in Example 3.4 to justify such a split). Hence, based on the inductive assumption, the production $A \to bAA$ allows to generate it.

  Based on the inductive assumption we can see that there is no way to derive from A any word of length $2n + 1$ with numbers of as and bs different than $n + 1$ and n,
- likewise, we can show that any word of length $2n + 1$ having n symbols a and $n + 1$ symbols b are derivable from B, and only such words,

- any word of length $2n + 2$ having symbols a and $n + 1$ symbols b is derivable from S. Such a word
  - either begin with a and have n symbols a and $n + 1$ symbols b in the remaining part. Hence, the production $S \to aB$ allows to generate it,
  - or begin with b and have $n+1$ symbols a and n symbols b in the remaining part. In this case the production $S \to bA$ is employed to generate it.

  Based on the inductive assumption we can see that only word of length $2n + 1$ with equal numbers of as and bs can be derived form S.

The proof is complete.

5. The formula $L = \{w \in \{a,b\}^* : \#_a w = 2 * \#_b w > 0\}$ defines the language. The reader can prove that the following grammar generates it: $G = (\{S, A, B\}\{a, b\}, P, S)$ with the following set P of productions:

$$\begin{aligned} \text{P:}\quad & S \to aAB \mid aBA \mid bAA \\ & A \to a \mid aS \mid bAAA \\ & B \to b \mid bS \mid aaBB \mid abABB \mid abBAB \mid abBBA \end{aligned}$$

Hint: employ mathematical induction to prove correctness of the grammar, consider the function $\mu(w) = \#_w a - 2 * \#_w b$.

Problem 3.2. Prove that the language L is context-free.

$$L = \{w \in \{a,b,c\}^* : w = a^k b^k c^l,\ k, l > 0\}.$$

Solution. The following grammar generates this language: $G = (\{S, A, B\}\{a, b\}, P, S)$ with the following set P of productions:

$$\begin{aligned} \text{P:}\quad & S \to AC \\ & A \to aAb \mid ab \\ & C \to Cc \mid c \end{aligned}$$

Problem 3.3. Prove that the language L is not context-free.

$$L = \{w \in \{a,b,c\}^* : w = a^k b^k c^k,\ k > 0\}.$$

Solution. *Hint*: apply contrapositive of the pumping lemma.

Problem 3.4. Prove that the language L is not context-free.

$$L = \{w \in \{a,b,c\}^* : \#_a w = \#_b w = \#_c w > 0\}$$

Solution. We apply the contrapositive of the pumping lemma to prove that L is not context-free. Let N is a constant from the lemma and let $z = a^N b^N c^N \in L$. We consider any split $z = uvwxy$ satisfying premises of the lemma, that is, $|vwx| \le N$ and $|vx| \ge 1$.

Thus, vwx cannot include three different symbols. As a result, vx is a string that includes at least one symbol of the alphabet and does not include at least one symbol, that is, it is either a string of as or a string of bs or a string of cs, or it is a string of as and bs or a string of bs and cs and it is never a string of as and bs and cs. In the word $z_0 = uv^0wx^0y$ number(s) of the symbol(s), which are present in vx, is decreased while number(s) of the symbol(s), which are absent in vx is not changed comparing to the word z. So then, $z_0 \notin L$, what ends the proof.

Problem 3.5. Prove that the language L is context-free.

$$L = \{wuw^R v : u, v, w \in \{a, b\}^*,\ |w| > 0\}.$$

Problem 3.6. Prove that the language $L = \{ww : w \in \{a, b\}^*\}$ is not context-free.

Solution. Apply contrapositive of the pumping lemma. Consider the word $z = a^N b^N a^N b^N$, where N is a natural number not less than the constant from the lemma.

Problem 3.7. Prove that complement of the language considered in Problem 3.6, that is $L = \{a, b\}^* - \{ww : w \in \{a, b\}^*\}$, is a context free one.

Solution. A context-free grammar generating the language L proves that this language is context-free. Words of odd length are not of a form ww, so they belong to L. The following productions generate such words:

$$\begin{aligned} \text{P}': \quad & S \to aR \mid bR \\ & R \to aaR \mid abR \mid baR \mid bbR \mid \varepsilon \end{aligned}$$

Let us consider words of even length. Any word of L is of a form

$$w = a_1\, a_2 \dots \underline{a_i} \dots a_n\, a_{n+1}\, a_{n+2} \dots \underline{a_{n+i}} \dots a_{n+n}$$

where $a_i \neq a_{n+i}$ for some $1 \leq i \leq n$ and no more restrictions are put on symbols of such a word. This word can be rewritten as:

$$a_{i-(i-1)} a_{i-(i-2)} \cdots a_{i-1} \underline{a_i}\, a_{i+1} \cdots a_{i+(i-1)}\, a_{(n+i)-(n-i)} \cdots a_{(n+i)-1} \underline{a_{n+i}}\, a_{(n+i)+1} \cdots a_{(n+i)+(n-i)}$$

As shown, every indicated symbol has a left context and a right context, which are any string of the same length for a given symbol.

Now, we can design a context-free grammar generating such words:

$$G = (\{S, A, B, C\}, \{a, b\}, P, S)$$

$$\begin{aligned} \text{P}'': \quad & S \to AB \mid BA \\ & A \to a \mid CAC \\ & B \to b \mid CBC \\ & C \to a \mid b \end{aligned}$$

Finally, joining bot sets of productions P′ and P″, we get productions of a context-free grammar generating this language:

$$\begin{aligned} \text{P:}\quad & S \to AB \mid BA \mid aR \mid bR \\ & A \to a \mid CAC \\ & B \to b \mid CBC \\ & C \to a \mid b \\ & R \to aaR \mid abR \mid baR \mid bbR \mid \varepsilon \end{aligned}$$

The detailed justification that the grammar G generates the language L is left to the reader.

Also, the reader can prove that the language L is not regular.

Hint: consider the Problem 2.5.

Problem 3.8. A language L is a context-free one. Is L' context-free?

$$L' = \{a_1\, a_2\, a_2\, a_3\, a_3\, a_3\, a_4\, a_4\, a_4\, a_4 \dots (a_n)^n : a_1\, a_2\, a_3\, a_4 \dots a_n \in L\}.$$

Solution. Let apply contrapositive of the pumping lemma. Let N is a natural number not less than the constant from the lemma such that there is a word $z = a_1\, a_2 \dots a_N \in L$. Take the word $z' = a_1\, a_2\, a_2\, a_3 \dots (a_N)^N$, which satisfies assumption of the lemma. Notice that $|z'| = N(N+1)/2$. Let $z' = uvwxy$ is a split holding the lemma assumptions. A word $z_2' = uv^2wx^2y$ has length $|z_2'| = N(N+1)/2 + r$ such that $1 \le r \le N$. Therefore, $N(N+1)/2 < |z_2'| = N(N+1)/2 + r \le N(N+1)/2 + N < (N+1)(N+2)/2 = N(N+1)/2 + N + 1$. But words of such length do not belong to L'. This proves that L' is not context-free.

4 Context-sensitive grammars and unrestricted grammars

The class of context-sensitive languages follows the class of context-free languages. In the hierarchy of languages, a so-called Chomsky hierarchy, the next classes of languages, besides regular languages and context-free languages, are context-sensitive and recursively enumerable languages. Context-sensitive languages are generated by context-sensitive grammars, which happen to be a generalization of context-free grammars. Recursively enumerable languages are generated by unrestricted grammars, which are an extension of context-sensitive grammars. We will also distinguish the class of recursive languages. However, a class of grammars generating recursive languages is not known. The class of recursive languages is separated from the class of recursively enumerable based on special class of automata. This topic will be presented in Chapter 5.

As mentioned above, context-sensitive and unrestricted grammars are much more complex than context-free and regular grammars. Likewise, the structure of words of these classes of languages is much more complex than the structure of words of context-free languages. We neither know properties exhibiting restriction of words' structure nor can we formulate any regularity rules of a language structure. Also, we do not know any property like, for instance, pumping lemmas, which would suggest a character analogous to the finiteness of regular and context-free languages. Algebraic characterization of these languages, as a substructure in the set of all words, is not known. Therefore, we do not have effective tools for processing these languages, as it is in cases of simpler classes of languages. Tools corresponding to pumping lemmas, Myhill–Nerode Theorem, CYK algorithm, etc., are not known for classes of context-sensitive and recursively enumerable languages.

This Chapter provides a short presentation of the basic properties of context-sensitive and recursively enumerable languages.

4.1 Context-sensitive grammars

Productions of *context-sensitive grammars* satisfy *monotonic condition* (also called *non-contracting* or *non-erasing condition*). For this reason, context-sensitive grammars are also called *monotonic, non-contracting* or *non-erasing grammars*.

Definition 4.1. A grammar $G = (V, T, P, S)$ is context-sensitive if and only if its productions are monotonic (or noncontracting), that is, they are of the form:

$$\alpha \to \beta, \quad \text{where: } \alpha, \beta \in (V \cup T)^* \text{ and } 0 < |\alpha| \leq |\beta|,$$

where $|w|$ denotes a length of the word w.

https://doi.org/10.1515/9783112223123-005

Definition 4.2. Context-sensitive languages are those generated by context-sensitive grammars and only those.

The class of context-sensitive grammars and the class of context-sensitive languages are denoted by CSG and CSL, respectively.

The monotonic condition excludes the empty word ε from context-sensitive languages. In terms of the above definition, any language, which includes the empty word, is not context-sensitive. This is the strict meaning of classes of CSL and CSG.

It would be unreasonable to exclude from the CSL class a context-sensitive language with the empty word included. Thus, any language L, such that $L - \{\varepsilon\}$ is a context-sensitive language, will be included in the CSL class. Note that having a language L generated by a monotonic grammar, we can add the empty word ε to L by attaching the production $S \rightarrow \varepsilon$ to the grammar, where S is the initial symbol of the grammar. This production breaks the monotonicity of the grammar, so it will be considered to be the unique exception of context-sensitive grammar. This meaning of context-sensitivity is called extensive context-sensitivity.

Summarizing the above notes: the classes of CSL and CSG will be considered in the strict or extended sense depending on the context of the discussion. In the sequel, we will not distinguish between strictness and extensiveness of context-sensitivity if this does not lead to confusion.

Definition 4.3. A grammar $G = (V, T, P, S)$ is in context-sensitive normal form if and only if its productions have the following form:

$$\gamma A \delta \rightarrow \gamma \alpha \delta, \quad \text{where } A \in V,\ \alpha, \gamma, \delta \in (V \cup T)^*,\ \alpha \neq \varepsilon.$$

γ and δ are called left and right context, respectively and $A \rightarrow \alpha$ is called the core of the production.

Note that a grammar in context-sensitive normal form is a context-sensitive grammar since its productions are monotonic. Then the class of grammars in context-sensitive normal form is included in the CSG class. The question is whether the CSG class is equivalent to the class of grammars in context-sensitive normal form. This question is equivalent to the question if, for any context-sensitive grammar, we can find an equivalent grammar in context-sensitive normal form. The answer is affirmative. Hence, grammars in context-sensitive normal form do not create a new class of languages; they generate the CSL class.

Note that the core of a context-sensitive production in normal form is simply a context-free production. Also observe, that only the core of such a production can affect derivation. However, the core of a production can be used only if left and right contexts are preserved. This is why grammars with monotonic productions are called context-sensitive grammars.

Now, let us justify that every context-sensitive grammar can be transformed to normal form:

Lemma 4.1. *Any context-sensitive grammar $G = (V, T, P, S)$ can be transformed to an equivalent grammar in normal form. That is, both grammars generate the same language.*

Proof. We build a context-sensitive grammar in the normal form, which is equivalent to the grammar G. An idea of proof of equivalence of both grammars is illustrated in Example 4.1. Below, an idea of such proof is outlined while details of proof is left to the reader.

The grammar in a normal form equivalent to a given context-sensitive grammar is designed as follows:

1. for every terminal symbol $a \in T$
   a. create a new non-terminal symbol A_a and convert every production of the grammar G replacing every occurrence of the symbol a with the new non-terminal A_a,
   b. add the new production $A_a \to a$.

   We get a new grammar $G_T = (V \cup V_T, T, P' \cup P_T, S)$ with an extended set $V \cup V_T$ of nonterminal symbols and an extended set $P' \cup P_T$ of productions, where $V_T = \{A_a : a \in T\}$, P' is the set of productions converted from P, $P_T = \{A_a \to a,\ a \in T\}$. Productions of P' have now a form $A_1 A_2 \dots A_k \to B_1 B_2 \dots B_l$, where $k \leq l$ (since the grammar G is a context-sensitive, that is, it is a monotonic grammar) and all symbols in the production are nonterminal symbols, that is, $A_1, \dots, A_k, B_1, \dots, B_l \in V \cup V_T$,
2. let us split the set P' of productions to subsets P'_n (corresponding to the subset P_n of P) of productions in normal form and P'_m (corresponding to the subset P_m of P) of productions that are monotonic, but not in normal form, $P = P_n \cup P_m$ and $P' = P'_n \cup P'_m$. Let us enumerate productions of the set P'_m, which are not in normal form,
3. for each production $r : A_1 A_2 \dots A_k \to B_1 B_2 \dots B_l$ in the set P'_m with assigned number r do:
   a. create a new nonterminal symbols A^r_k,
   b. replace the production $r : A_1 A_2 \dots A_k \to B_1 B_2 \dots B_l$ with the following set R_r of $k + 1$ productions in normal form:
      - $A_1 \dots A_{k-1} A_k \to A_1 \dots A_{k-1} A^r_k$, where $\gamma = A_1 \dots A_{k-1}$ is the left context, the right context is empty and $A_k \to A^r_k$ is the core,
      - $A_1 A_2 \dots A_{k-1} A^r_k \to B_1 A_2 \dots A_{k-1} A^r_k$, where the left context is empty, $\delta = A_2 A_3 \dots A_{k-1} A^r_k$ is the right context and $A_1 \to B_1$ is the core,
      - $B_1 A_2 A_3 \dots A_{k-1} A^r_k \to B_1 B_2 A_3 \dots A_{k-1} A^r_k$, where $\gamma = A_1$ is the left context, $\delta = A_3 \dots A_{k-1} A^r_k$ is the right context and $A_2 \to B_2$ is the core,
      - ...
      - $B_1 \dots B_{k-2} A_{k-1} A^r_k \to B_1 \dots B_{k-2} B_{k-1} A^r_k$, where $\gamma = B_1 \dots B_{k-2}$ is the left context, $\delta = A^r_k$ is the right context and $A_{k-1} \to B_{k-1}$ is the core,
      - $B_1 \dots B_{k-1} A^r_k \to B_a \dots B_{k-1} B_k \dots B_l$, where $\gamma = B_1 \dots B_{k-1}$ is the left context, the right context is empty and $A^r_k \to B_k \dots B_l$ is the core,

4. finally we come up with the following context-sensitive grammar in normal form $G_N = (V_N, T, P_N, S)$, where:
   - $V_N = V \cup V_T \cup \bigcup_r \{A^r_{k_r}\}$ and r is the number of a production $A_1 A_2 \ldots A_k \to B_1 B_2 \ldots B_l$ from P'_m,
   - $P_N = P_T \cup P_n \cup \bigcup_r R_r$, where R_r is the set of productions in normal form corresponding to the production r of P_m. □

Example 4.1. Let us consider the context-sensitive grammar $G = (\{S\}, \{a, b, c\}, P, S)$ with the following productions:

$$
\begin{array}{lrl}
\text{P:} & S \to abcS \mid abc & (1)(2) \\
& ab \to ba & (3) \\
& ac \to ca & (4) \\
& ba \to ab & (5) \\
& bc \to cb & (6) \\
& ca \to ac & (7) \\
& cb \to bc & (8)
\end{array}
$$

This is a context-sensitive grammar generating the set of words with the same number of letters a, b and c, that is, the language $L = \{w : w \in \{a, b, c\}^* \text{ and } \#_a w = \#_b w = \#_c w > 0\}$. The proof is straightforward: productions (1) and (2) insert any number of letters a, b and c (of course, the same number of letters a, b and c). Remaining productions allow for migration of any letter to any place in the word, so any permutation of letters generated by productions (1) and (2) is available.

The grammar G is not in the normal form. Let us transform it to the normal form according to Lemma 4.1.

1. replacement of terminal symbols by new non-terminals and extension of the set of productions leads to the following grammar

$$G' = (\{S, A, B, C\}, \{a, b, c\}, P' \cup P_T, S)$$

$$
\begin{array}{lrl}
\text{P}': & S \to ABCS \mid ABC & (1)(2) \\
& AB \to BA & (3) \\
& AC \to CA & (4) \\
& BA \to AB & (5) \\
& BC \to CB & (6) \\
& CA \to AC & (7) \\
& CB \to BC & (8) \\
P_T: & A \to a & \\
& B \to b & \\
& C \to c &
\end{array}
$$

2. for productions (3–8) of the grammar G', which are not in normal form, create the set P' corresponding to the set P of original productions 3–8 of the grammar G (we keep their numbers),
3. the set P' of productions is processed as follows:
   a. new non-terminals are created: $B^3, C^4, A^5, C^6, A^7, B^8$,
   b. every production of the set P'_m is replaced as follows:
      – the production 3: $AB \to BA$ is replaced by the following set of productions: $3_N : AB \to AB^3,\ AB^3 \to BB^3,\ BB^3 \to BA$,
      – the production 4: $AC \to CA$ is replaced by the following set of productions: $4_N : AC \to AC^4,\ AC^4 \to CC^4,\ CC^4 \to CA$,
      – etc.,
4. finally, we build the following context-sensitive grammar in normal form $G_N = (V_N, T, P_N, S)$, where:
   – $V_N = \{S, A, B, C, B^3, C^4, A^5, C^6, A^7, B^8\}$,
   –

$$
\begin{array}{llrcll}
P_N: & & S & \to & ABCS \mid ABC & (1)(2) \\
& & AB & \to & AB^3 & (3) \\
& & AB^3 & \to & BB^3 & \\
& & BB^3 & \to & BA & \\
& & AC & \to & AC^4 & (4) \\
& & AC^4 & \to & CC^4 & \\
& & CC^4 & \to & CA & \\
& & BA & \to & BA^5 & (5) \\
& & BA^5 & \to & AA^5 & \\
& & AA^5 & \to & AB & \\
& & BC & \to & BC^6 & (6) \\
& & BC^6 & \to & CC^6 & \\
& & CC^6 & \to & CB & \\
& & CA & \to & CA^7 & (7) \\
& & CA^7 & \to & AA^7 & \\
& & AA^7 & \to & AC & \\
& & CB & \to & CB^8 & (8) \\
& & CB^8 & \to & BB^8 & \\
& & BB^8 & \to & BC & \\
P_T: & & A & \to & a & \\
& & B & \to & b & \\
& & C & \to & c &
\end{array}
$$

Both grammars G and G_N are equivalent. To prove this equivalence, it is sufficient to show that every word generated in one grammar is also generated in another one.

Having a derivation of a word $w \in \{a, b, c\}^*$ in the grammar G, we can turn it to a derivation of the same word in the grammar G_N:

- every step of the derivation using the production number r, $1 \le r \le 8$, of the set P is replaced by steps applying the corresponding set r of productions R_N of the grammar G_N,
- the new derivation generates the word $W \in \{A, B, C\}^*$ which reflects the word w. Now, applying productions of the set P_T to non-terminal symbols A, B, C we get the word w.

For example, the derivation in the grammar G:

$$\underline{S} \xrightarrow{1} \underline{ab}cS \xrightarrow{3} b\underline{acS} \xrightarrow{4} \underline{bc}aS \xrightarrow{6} cba\underline{S} \xrightarrow{2} cbaabc$$

could be transformed to the following derivation in the grammar G_N:

$$\begin{aligned}
&\underline{S} \xrightarrow{1} \underline{AB}CS \\
&\xrightarrow{3_N} \underline{AB^3}CS \xrightarrow{3_N} \underline{BB^3}CS \xrightarrow{3_N} B\underline{ACS} \\
&\xrightarrow{4_N} \underline{BAC^4}S \xrightarrow{4_N} \underline{BCC^4}S \xrightarrow{4_N} \underline{BC}AS \\
&\xrightarrow{6_N} \underline{BC^6}AS \xrightarrow{6_N} \underline{CC^6}AS \xrightarrow{6_N} CBA\underline{S} \\
&\xrightarrow{2} \underline{C}BAABC \to c\underline{B}AABC \to cb\underline{A}ABC \xrightarrow{*} cbaabc
\end{aligned}$$

In the above derivation, the number of production (or set of productions) is indicated above the arrow. A part of the intermediate word used in a production is underlined.

On the other hand, a derivation in the grammar G_N can be transformed to a derivation in the grammar G:

- turn a derivation in the grammar G_N into derivation in the grammar G':
  - notice that if any production from groups 3–8 of P_N is used, all three productions of this group must appear consecutively one after another. For instance, any production of the 3rd group must appear in the following sequence:

$$\cdots \to \ldots \underline{AB} \ldots \xrightarrow{3_N} \ldots \underline{AB^3} \ldots \xrightarrow{3_N} \ldots \underline{BB^3} \ldots \xrightarrow{3_N} \ldots BA \ldots \to \cdots$$

  - replace any triad as indicated above by the corresponding production from the set P',
- turn the above derivation in the grammar G_T into derivation in the grammar G,
  - replace any step of derivation applying a production $A_a \to a$ by the terminal symbol a. For instance: replace $cdots \to \alpha \underline{A} \beta \to \alpha a \beta \to \cdots$ by $\cdots \to \alpha a \beta \to \cdots$

Notice that application of any production from the set R_N must start from the first production of the given triad number r. It inserts the new nonterminal symbol A_k^r, which is included in the left-hand side of the remaining productions of this set. Then consecutive productions of the set R_N must be applied; otherwise, the nonterminal symbol A_k^r cannot be eliminated. Only such a sequence could be replaced by the production number r of the grammar G'. If the sequence of productions of the set R_N is broken, then either a terminal word cannot be derived, or a break is made in this sequence by a nested sequence of another set Q_N of productions. In the latter case, if the result of productions is nested in a left-hand side of a production R_N, then the production number q can be omitted. Otherwise, it should be used in derivation prior to the production number r.

4.2 Unrestricted grammars

The class of unrestricted grammars is the most general class of grammars. Unrestricted grammars are similar to context-sensitive grammars except that productions are not required to be monotonic (non-contracting). Unrestricted grammars generate the class of recursively enumerable languages, which will be denoted as REL class of languages. The formal definitions are as follows:

Definition 4.4. A grammar $G = \{V, T, P, S\}$ is unrestricted if and only if its productions are of the form:

$$\alpha \to \beta, \quad \text{where: } \alpha, \beta \in (V \cup T)^* \text{ and } 0 < |\alpha|$$

where: $|w|$ denotes length of the word w.

Definition 4.5. Recursively enumerable languages are those generated by unrestricted grammars and only those.

Context-sensitive grammars are special cases of unrestricted grammars. Thus, the class of context-sensitive languages is a subclass of recursively enumerable languages, that is, $CSL \subset REL$. But it is not obvious if this inclusion is proper, that is, if $CSL \neq REL$ because nearly every language that we can imagine is context-sensitive. In the consecutive Chapters, we will design languages that are recursively enumerable but not context-sensitive.

Example 4.2. The language $L = \{a^k b^k c^k d^k \ : \ k > 0\}$ is not context-free (we can apply contraposition of the pumping lemma for context-free languages to prove it). Design a context-sensitive grammar generating this language.

First, we design an unrestricted grammar generating this language. Notice that words of this language can be ordered according to their length:

$$L = \{abcd, aabbccdd, aaabbbcccddd, \ldots\}.$$

The first word in this order is *abcd*. Having a word of this language, we can design the next one (in this order) by supplementing: (i) with a the sequence of a's, (ii) with b the sequence of b's, (iii) with c the sequence of c's and (iv) with d the sequence of d's. Based on this observation, we will design a grammar in which consecutive words of the language will be designed by enlarging sequences of letters a, b, c, d. Enlargement will be done by symbols traveling along the word. The following grammar generates the language: $G = (\{S, A, D\}, \{a, b, c, d\}, P, S)$ with productions:

$$
\begin{array}{llr}
\text{P:} & S \to abcDd & (1) \\
& cD \to Dc & (2) \\
& bD \to Db & (3) \\
& aD \to aaA & (4) \\
& Ab \to bA & (5) \\
& bAc \to bbcA & (6) \\
& cAc \to ccA & (7) \\
& cAd \to ccDdd & (8) \\
& D \to \varepsilon & (9)
\end{array}
$$

A derivation of the word $w = aaabbbcccddd$ is as follows:

$$
\begin{aligned}
&\underline{S} \xrightarrow{1} ab\underline{cD}d \xrightarrow{2} a\underline{bD}cd \xrightarrow{3} \underline{aD}bcd \xrightarrow{4} aa\underline{Ab}cd \xrightarrow{5} aa\underline{bAc}d \xrightarrow{6} aabb\underline{cAd} \\
&\xrightarrow{8} aabbc\underline{cD}dd \xrightarrow{2*2} aab\underline{bD}ccdd \xrightarrow{2*3} a\underline{aD}bbccdd \xrightarrow{4} aaa\underline{Ab}bccdd \xrightarrow{2*5} aaa\underline{bAc}cdd \\
&\xrightarrow{6} aaabbb\underline{cAc}dd \xrightarrow{7} aaabbbc\underline{cAd}d \xrightarrow{8} aaabbbccc\underline{D}ddd \xrightarrow{9} aaabbbcccddd
\end{aligned}
$$

In the above derivation, the symbol $\xrightarrow{2*3}$ denotes an application of the production no. 3 two times.

The derivation is terminated if and only if production no. 9 is applied. This production may be used for any placement of the non-terminal symbol D in an intermediate word of derivation. Note that production no. 9 can be applied only to the non-terminal symbol D. If this symbol appears in an intermediate word of derivation, then this word includes the same number of letters a, b, c and d, so then removal of this symbol produces a word of the language L.

The above grammar is not context-sensitive because production no. 9 is nullable (not monotonic). On the other hand, the empty word ε is not generated in this grammar. Therefore, perhaps the grammar may be turned (though no evidence is given for a general case) to a context-sensitive one by eliminating the nullable production no. 9. Elimination of the nullable production is done here by grouping symbols cDd and substituting them by a new non-terminal symbol (denoted by $[cDd]$ to keep transformation clear). The following monotonic grammar generates the above language:

$$G = (\{S, A, D, [cDd]\}, \{a, b, c, d\}, P, S)$$

$$\begin{array}{lrcll}
\text{P:} & S & \rightarrow & ab[cDd] & (1)\\
& cD & \rightarrow & Dc & (2)\\
& bD & \rightarrow & Db & (3)\\
& aD & \rightarrow & aaA & (4)\\
& Ab & \rightarrow & bA & (5)\\
& bAc & \rightarrow & bbcA & (6)\\
& cAc & \rightarrow & ccA & (7)\\
& cAd & \rightarrow & c[cDd]d & (8)\\
& [cDd] & \rightarrow & cDd & (9)\\
& [cDd] & \rightarrow & cd & (10)
\end{array}$$

A derivation of the word $w = aaabbbcccddd$ looks now as follows:

$$\underline{S} \overset{1}{\rightarrow} ab\underline{[cDd]} \overset{9}{\rightarrow} ab\underline{cD}d \overset{2}{\rightarrow} a\underline{bD}cd \overset{3}{\rightarrow} \underline{aD}bcd \overset{4}{\rightarrow} aa\underline{Ab}cd \overset{5}{\rightarrow} aa\underline{bAc}d$$
$$\overset{6}{\rightarrow} aabb\underline{cAd} \overset{8}{\rightarrow} aabbc\underline{[cDd]}d \overset{9}{\rightarrow} aabbc\underline{cD}dd \overset{2*2}{\rightarrow} aab\underline{bD}ccdd$$
$$\overset{2*3}{\rightarrow} a\underline{aD}bbccdd \overset{4}{\rightarrow} aaa\underline{Ab}bccdd \overset{2*5}{\rightarrow} aaa\underline{bAc}cdd \overset{6}{\rightarrow} aaabbb\underline{cAc}dd$$
$$\overset{7}{\rightarrow} aaabbbc\underline{cAd}d \overset{8}{\rightarrow} aaabbbcc\underline{[cDd]}dd \overset{10}{\rightarrow} aaabbbcccddd$$

Part II: **Automata and accepting languages**

5 Turing machines

In Part I, the methods of generating languages were studied. Those methods are based on different types of grammars and on regular expressions. Part II is devoted to a discussion on methods of languages acceptance. Acceptation of languages is based on different types of automata: Turing machines, linear bounded automata, pushdown automata and finite automata. Identification of grammars with the generation of languages and automata with the acceptation of languages is a subjective and intuitive categorization done by the authors. However, it reflects the nature of tools for processing languages.

Turing machines (and other types of automata) can be interpreted as models of computation. Turing machines is a universal model of computation that is used for such purposes as, for instance, acceptance of languages, computing functions, solving problems.

In this Chapter, Turing machines, and automata in general, will be employed as tools of acceptance of languages, that is, they will be queried whether a given the word is in the language accepted by a given automaton or not.

Turing machines can also compute functions. Such machines compute functions with natural numbers as domain and co-domain. Another type of Turing machines solves problems like, for instance, the sorting problem, the shortest paths problem, etc. We only touch these aspects here.

5.1 Deterministic Turing machines

In this book, two categories of *Turing machines* (say *automata*, in general) will be studied: *deterministic* and *nondeterministic*. Roughly speaking, the computation of an automaton is a sequence of configurations organized according to some control information. An automaton is a deterministic one if and only if there is at most one possibility of doing a transition in any configuration. If, for a given automaton, there is a choice of doing a transition in some configuration(s), then such an automaton is a nondeterministic one.

In this section, different categories of deterministic Turing machines are studied: basic model, model with the guard, multi-track model and multi-tape model. At the end of this Chapter nondeterministic Turing machines are discussed. Equivalence of these categories of Turing machines is drawn, that is, it is shown that for a Turing machine in any model, we can find an equivalent machine in any other model. Equivalence of Turing machines (equivalence of automata, in general) means that they accept the same language, compute the same function or solve the same problem. The discussion leads to the main goal of this Chapter, that is, that the class of deterministic Turing machines and the class of nondeterministic Turing machines are equivalent.

https://doi.org/10.1515/9783112223123-007

5.1.1 Basic model of Turing machines

The definition of basic model of deterministic Turing machines is given below. Later in this Chapter, other deterministic models of Turing machines are discussed. They are proved to be equivalent to basic model. As it was stated above, equivalence with regard to accepted languages is considered. Besides, generalization of equivalence issue to Turing machines computing functions or solving problems is straightforward.

Definition 5.1. A Turing machine in basic model is a system

$$M = (Q, \Sigma, \Gamma, \delta, q_0, B, F, C)$$

with components as follows:
Q a finite set of states,
Γ a finite set of tape symbols (tape alphabet),
B the blank symbol (of tape alphabet), $B \in \Gamma$,
Σ an input alphabet, $\Sigma \subset (\Gamma - \{B\})$,
q_0 the initial state, $q_0 \in Q$,
F a set of accepting states, $F \subset Q$,
C a condition, its satisfaction is necessary and sufficient to stop computation,
δ a transition function, which is a mapping: $\delta : Q \times \Gamma \to Q \times \Gamma \times \{L, R\}$

where L, R denote left and right directions.

Notice that a transition function δ may not be a total function, that is, it may be undefined for some of its arguments. Such a case is formally interpreted that the machine falls into an infinite computation. This comment will be explained in detail in the further discussion of this Chapter.

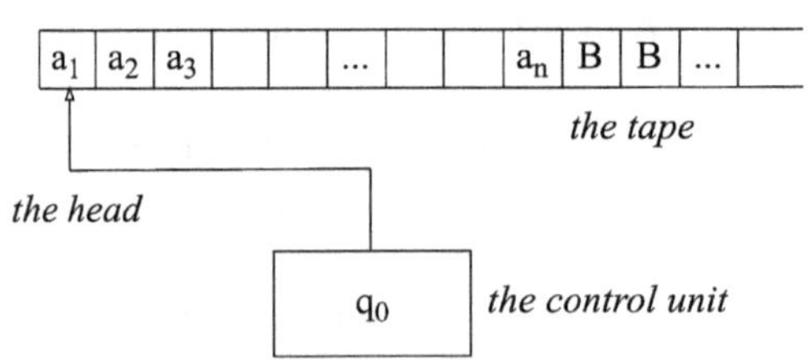

Figure 5.1: Basic model of Turing machine.

A Turing machine could be interpreted as a physical mechanism shown in Figure 5.1. This mechanism consists of:
- a control unit, it is in a state of Q,
- a one-way infinite tape split into cells; every cell contain a symbol (exactly one) of the tape alphabet Γ,

- the head, it is placed over a cell of a tape, it reads a symbol held in a cell, it stores the desired symbol in the cell, it shifts left or right.

Turing machines do the computation for a given input data. The computation of a given Turing machine is done according to the following intuitive procedure:

1. the initial configuration of a given Turing machine is described as follows:
   a. input data, a word $w = a_1 a_2 \dots a_n$ over input alphabet Σ, is stored in n beginning cells of the tape, cf. Figure 5.1,
   b. all other cells of the tape, which is infinite to the right, are filled in with the blank symbol B,
   c. the head of the Turing machine is placed over the first (leftmost) cell of the tape,
   d. the control unit is in the beginning state q_0,
2. if the stop condition C is satisfied, then the computation is halted, the configuration is called the final configuration, the machine responses whether its control unit is in an accepting state or not,
3. if the stop condition C is not satisfied, then – based on the state q of the control unit and the symbol X read by the head – the Turing machine is doing the following actions:
   a. the value (p, Y, D) of the transition function $\delta(q, X)$ is computed,
   b. the head stores the tape symbol Y in the cell under it,
   c. the control unit switches to the state p,
   d. the head shifts by one cell in the direction D,
4. computation goes to the point 2.

The above intuitive procedure could be adapted to Turing machines, which compute functions or solve problems. This adaptation needs a redefinition of input data. Input data of a Turing machine computing function or solving problem is:

- a sequence of arguments of the function computed by the machine. Arguments of a function are encoded as numbers, e. g., in the binary or decimal positional system or in the unary system. Of course, some sort of separators between arguments must be used,
- a data defining an instance of the problem solved by the machine. This data is encoded in a way depending on the type of data.

Note that the stop condition C may never be satisfied and the machine will be doing an infinite computation. When the infinite computation is done, the input data is not accepted by the machine, that is, in the case of accepting a language, the input word does not belong to the language accepted by the Turing machine. It is worth underlining that a Turing machine always stops its computation in an accepting state if and only if its input data is a word of the language accepted by the machine. Turing machines raise a difficult problem: when one is performing long computation, there is no indication if the machine has fallen into infinite computation or it will stop computation in the future.

Remark 5.1. For the sake of clarity, we assume that Turing machines will be designed assuming that, if a machine terminates computation, its head is placed over the first (leftmost) cell and:

- when the machine accepts a language, then all cells of the taped are filled in with the blank symbol B,
- when the machine computes a function, then the value of the function is stored in the beginning cells of the tape. All other cells are filled in with the blank symbol B. The stored value is the correct result if and only if the machine stopped computation in an accepting state,
- when the machine solves a problem, then output data is stored in the beginning cells of the tape. All other cells of the tape are filled in with the blank symbol B. Output data is a correct solution of a computed instance of the problem if and only if the machine stops computation in an accepting state.

Assumptions of the above remark are not included in Definition 5.1 and are not necessary, but they are desired for clarity of computation. An epilogue of a computation guarantying satisfaction of the above assumptions is called a cleaning procedure.

Definition 5.2. A *step description* (a *configuration*) of a Turing machine

$$M = (Q, \Gamma, \Sigma, \delta, q_0, B, F, C)$$

is a sequence of symbols:

$$\alpha_1 \, q \, \alpha_2$$

where:

- $q \in Q$ is the current state of the control unit of a Turing machine,
- α_1 is a sequence of symbols stored in cells beginning from the leftmost one and ending with the cell prior to the one under the head,
- α_2 is a sequence of symbols stored in cells beginning from the one under the head, going to the right and ending with the rightmost one holding a non-blank symbol.

Note that both α_1 and α_2 sequences of symbols are words over the tape alphabet Γ and that any of these sequences may be the empty word. However, none of these two sequences can be infinite. This is due to the following reasons:

- an input data is finite, so then only a finite number of cells are filled in with non-blank symbols in the input configuration,
- after any step of computation, only a finite number of cells could be visited by the head. Therefore, only a finite number of cells may get a nonempty symbol.

For instance, the initial step description (configuration) is of the form $q_0 \, w$, where q_0 is the initial state, w is the input data. In this case, α_1 is the empty word. On the other hand, a step description $\alpha_1\, q$ informs that all cells from the one under the head to the right are filled in with the blank symbol B. Now, α_2 is the empty word. Finally, when a Turing machine accepting a language ends its computation in a state q, then – according to Remark 5.1 – q will be the final step description.

Let us analyze transitions done by Turing machines. We assume that a step description is characterized by the following sequence of symbols:

$$X_1\, X_2\, \ldots\, X_{i-1}\, q\, X_i\, \ldots\, X_n$$

Recall that:
- q is the current state of the control unit,
- $X_1\, X_2\, \ldots\, X_{i-1}$ is the sequence of symbols of the tape alphabet Γ stored in cells preceding the cell under the head,
- $X_i\, X_{i+1}\, \ldots\, X_n$ is the sequence of symbols stored in cells beginning from the one under the head, going to the right and ending with the rightmost one containing a non-blank symbol,
- the head is placed over the cell with the X_i symbol stored in. If the sequence $X_i\, X_{i+1}\, \ldots\, X_n$ is the empty word, then the head reads the blank symbol B.

The transition function determines the following step description:
- if the value of the transition function is $\delta(q, X_i) = (p, Y, R)$, that is, the control unit switches to the state p, the head stores Y and shifts right, then we get the following configuration,

$$X_1\, X_2\, \ldots\, X_{i-1}\, Y\, p\, X_{i+1}\, \ldots\, X_n$$

- if the value of the transition function is $\delta(q, X_i) = (p, Y, L)$, then we get the following configuration:

$$X_1\, X_2\, \ldots\, X_{i-2}\, p\, X_{i-1}\, Y\, X_{i+1}\, \ldots\, X_n$$

We will use the symbol $\succ$ to denote a transition of a Turing machine. The transition symbol $\succ$ may be supplement with a Turing machine name $\succ_M$ to emphasize that a transition concerns a given Turing machine. It also can be supplemented with a superscript $\succ^k$ to notify k transitions done.

The above two transitions done by a Turing machine will be denoted as follows:

$$X_1\, X_2\, \ldots\, X_{i-1}\, q\, X_i\, \ldots\, X_n \succ X_1\, X_2\, \ldots\, X_{i-1}\, Y\, p\, X_{i+1}\, \ldots\, X_n$$
$$X_1\, X_2\, \ldots\, X_{i-1}\, q\, X_i\, \ldots\, X_n \succ X_1\, X_2\, \ldots\, X_{i-2}\, p\, X_{i-1}\, Y\, X_{i+1}\, \ldots\, X_n$$

Definition 5.3. Transitions of a Turing machine create a binary relation in the space of all possible configurations of the machine, that is, any two configurations are related if and only if the second one is derived from the first one utilizing the transition function.

This relation is called the transition relation of a given Turing machine and is marked with the symbol $\succ$. We will also consider the transitive closure of the transition relation and mark it with the symbol $\succ^*$.

Definition 5.4. A computation of a Turing machine $M = (Q, \Gamma, \Sigma, \delta, q_0, B, F, C)$ is a sequence of configurations $\eta_1, \eta_2, \ldots, \eta_n$ such that η_1 is the initial configuration, η_n is the final configuration and a pair of any two successive configurations belongs to the transition relation. If the machine has fallen into an infinite computation, then its computation is an infinite sequence of configurations $\eta_1, \eta_2, \eta_3 \ldots$ such that η_1 is the initial configuration and any pair of two successive configurations belongs to the transition relation. A finite computations is denoted as $\eta_1 \succ \eta_2 \succ \cdots \succ \eta_n$ and infinite computation is denoted as $\eta_1 \succ \eta_2 \succ \eta_3 \succ \cdots$.

Now we give a formal definition of acceptation of an input by a Turing machine:

Definition 5.5. A Turing machine accepts its input if and only if the computation terminates in an accepting state. In other words, a Turing machine accepts its input if and only if the pair of the initial configuration and the final configuration belongs to the transitive closure of the transition relation, that is, $\eta_1 \succ^* \eta_T$, where η_1 is an initial configuration and η_T is a final (that is, accepting) configuration.

Based on the above discussion, we now give formal definitions of some concepts.

Definition 5.6. The language $L(M)$ accepted by a Turing machine M is the set of words $w \in \Sigma^*$ accepted by the Turing machine.

Definition 5.7. A function computed by a Turing machine M is a mapping from a space of input data into a space of output data. If the machine accepts its input, then its output is the correct value of the function. Otherwise, when the machine stops computation but not accepts the input, or if it is doing infinite computation, the function is undefined for such input data.

Remark 5.2. We assume that the blank symbols B will neither separate non-blank symbols on tape nor be placed in leftmost cells prior to a non-blank symbol. This assumption is not required by definitions and concepts discussed so far. However, it simplifies the designing of Turing machines for given tasks.

Example 5.1. Design a Turing machine computing the function:

$$f : N \to N, \quad f(n) = \left\lceil \frac{n}{3} \right\rceil, \qquad \text{where } N = \{0, 1, 2, \ldots\}$$

Solution. First, we briefly comment on an algorithm for a Turing machine computing this function. The algorithm consists of a method of encoding input and output data and of a description of computation:

- input and output data are stored in the unary system, that is, the number n is stored as the sequence of n unary digits 0,
- computation of the machine relies on repeated subtraction of the denominator from the numerator. Subtraction is done by removing three digits 0 from data stored on the tape:
  - marking the first digit 0, it increments the function value, that is, for every subtraction, the value of the function is incremented by 1,
  - deleting the last two zeros,
- in the last subtraction, there might be one or no digit 0 to delete,
- the function value will be equal to the number of subtractions, that is, to the number of marked digits 0.

A Turing machine computing this function is as follows:

$$M = (Q, \Sigma, \Gamma, \delta, q_0, B, F, C)$$

where:
- $Q = \{q_0, q_1, q_2, q_3, q_4, q_5, q_6, q_A\}$,
- $\Sigma = \{0\}$,
- $\Gamma = \{0, A, X, B\}$,
- $F = \{q_A\}$ and
- the stop condition C is reached if control unit switches to the accepting state q_A.

The transition function is given in Table 5.1.

Table 5.1: The transition function of the Turing machine of Example 5.1.

δ	0	A	X	B
q_0	(q_2, X, R)			(q_6, B, R)
q_1	(q_2, A, R)			(q_6, B, L)
q_2	$(q_2, 0, R)$			(q_3, B, L)
q_3	(q_4, B, L)	$(q_6, 0, L)$	$(q_6, 0, R)$	
q_4	(q_5, B, L)	$(q_6, 0, L)$	$(q_6, 0, R)$	
q_5	$(q_5, 0, L)$	(q_1, A, R)	(q_1, X, R)	
q_6	$(q_A, 0, L)$	$(q_6, 0, L)$	$(q_6, 0, R)$	(q_A, B, L)

The transition function is a computing program completed in a low-level programming language. The detailed description of the transition function is given here in the form of comments to states of the machine:

q_0 terminates computation for the input data (argument of the function) equal to zero or sets the output value to one for non-zero input data. The tape symbol X is stored in the first (leftmost) cell. This symbol indicates the output value as well as marks the leftmost cell of the tape,

q_1 increments output data by one for every subtraction. Incrementing is done by storing the tape symbol A in the leftmost cell filled in with the digit 0. If there are no more 0's, then a cleaning process begins with switching to the state q_6,

q_2 passes the head to the end of input data, that is, to the cell that directly follows the cell with the rightmost digit 0 and then goes back to the cell with the rightmost digit 0,

q_3 deletes the rightmost digit 0 and moves the head one cell left. If there are no more 0's, then a cleaning process begins with switching to the state q_6,

q_4 deletes the last but one 0, states q_0, q_3, q_4 or q_1, q_3, q_4 are in charge of removing three digits 0 from input data, that is they are in charge of subtraction 3 from the input data. If there are no more 0's, then a cleaning process begins with switching to the state q_6,

q_5 passes the head to the beginning of input data, that is, to the cell with the leftmost 0. Then the control unit switches to the state q_1 to repeat the process of subtraction of 3 from input data,

q_6 preparation to terminate computation passes the head to the beginning of input data and turns symbols A and X to 0's,

q_A terminates computation.

Now we provide computation for given input data *zero* and *five*. *Zero* is represented as the empty sequence of unary digits and *five* is represented as the sequence of 5 unary digits 00000:

- $f(0) = \lceil \frac{0}{3} \rceil = 0$:
  $q_0 \succ Bq_6 \succ q_A$
- $f(0) = \lceil \frac{5}{3} \rceil = 2$:
  $q_0 00000 \succ Xq_2\, 000 \succ X0q_2\, 000 \succ X00q_2\, 00 \succ X000q_2\, 0 \succ X0000q_2 \succ X000q_3\, 0 \succ X00q_4\, 0 \succ X0q_5\, 0 \succ Xq_5\, 00 \succ q_5\, X00 \succ Xq_1\, 00 \succ XAq_2\, 0 \succ XA0q_2 \succ XAq_3\, 0 \succ Xq_4\, A \succ q_6\, X0 \succ 0q_6\, 0 \succ q_A\, 00$

5.1.2 Turing machine with the stop property

As noticed before, some Turing machines can fall into an infinite computation. Some other will terminate their computation for any input data. This observation draws the definition of the subclass of Turing machines which always terminate their computation.

Definition 5.8. A Turing machine in basic model with the stop property is a system introduced in Definition 5.1

$$M = (Q, \Sigma, \Gamma, \delta, q_0, B, F, C)$$

and such that it terminates its computation for any input data.

Example 5.2. Design a Turing machine with the stop property computing the function exposed in Example 5.1.

Solution. The transition table of the Turing machine of Example 5.1, as shown in Table 5.1, has empty entries for some arguments. The empty entries mean that the transition function is undefined for such arguments. Such configuration of a Turing machine, in which a value of its transition function is undefined, is understood as falling into the infinite computation. In light of this interpretation, the Turing machine of Example 5.1 does not have the stop property. On the other hand, a configuration in which the transition function is undefined is never reached. Moreover, the machine terminates its computation for any input data. However, we want to keep the assumption that undefined transition function value starts infinite computation. So then, to solve the inconsistency, we will turn the machine of Example 5.1 to have the (formal) stop property.

A new state q_R is added to states of the machine of Example 5.1. The control unit switches to this state any time when the original transition function is undefined (a special symbol is stored in the cell and the head goes right because this shift is always possible). Computation is terminated if and only if the control unit switches to any of two states q_A or q_R. Therefore, the response of the machine depends only on two states: q_A as accepting state and q_R as rejecting (not accepting) one.

Note, due to the sake of simplicity and readability of the transition function, the cleaning procedure of Remark 5.1 is not applied when computation reaches the state q_R.

Table 5.2: An updated transition function of the Turing machine designed in Example 5.1.

δ	0	A	X	B
q_0	(q_2, X, R)	$(q_R, \$, R)$	$(q_R, \$, R)$	(q_6, B, R)
q_1	(q_2, A, R)	$(q_R, \$, R)$	$(q_R, \$, R)$	(q_6, B, L)
q_2	$(q_2, 0, R)$	$(q_R, \$, R)$	$(q_R, \$, R)$	(q_3, B, L)
q_3	(q_4, B, L)	$(q_6, 0, L)$	$(q_6, 0, R)$	$(q_R, \$, R)$
q_4	(q_5, B, L)	$(q_6, 0, L)$	$(q_6, 0, R)$	$(q_R, \$, R)$
q_5	$(q_5, 0, L)$	(q_1, A, R)	(q_1, X, R)	$(q_R, \$, R)$
q_6	$(q_A, 0, L)$	$(q_6, 0, L)$	$(q_6, 0, R)$	$(q_R, \$, R)$

It is worth underlining that a function or a language computed by a Turing machine with the stop property can also be computed by a Turing machine without stop property. Moreover, such the Turing machine without stop property can perform infinite computation for some input data.

Two Turing machines, one with the stop property and another one without the stop property, if compute the same function or accept the same language, must terminate computation in accepting states for the same input data. Both machines may also terminate computation in rejecting states for the same data. They may yield different outputs only for such input data, which is not accepted. In such a case, the machine with the stop property terminates its computation in a rejecting state and the second machine falls in infinite computation.

Definition 5.9. Turing machines are considered equivalent if and only if for the same input data they either terminate computation in accepting state or none of them does it.

5.1.3 Simplifying the stop condition

Now we will slightly change definitions of Turing machines in basic model to simplify the definition of termination of its computation.

Definition 5.10. Turing machine with the halting accepting state is a system:

$$M = (Q, \Sigma, \Gamma, \delta, q_0, B, F)$$

where:
- $F = \{q_A\}$ – there is only one accepting state q_A,
- the stop condition is satisfied if and only if the machine switches to the accepting state q_A,
- other components of the system are as described in Definition 5.1.

Proposition 5.1. *Turing machines with the halting accepting state are equivalent to Turing machines in basic model.*

Proof. First of all, a Turing machine $M = (Q, \Sigma, \Gamma, \delta, q_0, B, \{q_A\})$ with halting accepting state formally matches Definition 5.1.

On the other hand, a Turing machine in basic model $M = (Q, \Sigma, \Gamma, \delta, q_0, B, F, C)$ can be updated to a machine with halting accepting state by:
- adding new states $q_\#$ and q_A,
- doing two transitions: $(q_\#, X, R)$ (q_A, Y, L), when the stop condition of the Turing machine in basic model is satisfied and the machine is in accepting state, where X and Y are symbols previously stored in cells under the head. These two transitions just switch the machine to the new state q_A, keeping contents of the tape unaffected and places the head in the same position as before these transitions,
- changing status of former accepting states to not accepting and assuming q_A as the only accepting state,
- redefining the stop condition: computation is halted if and only if the machine switches to the state q_A (now the only accepting state).

The machine with a halting accepting state may fall into infinite computation when the machine in basic model stops its computation in a non-accepting state. Anyway, the machine with the halting accepting state terminates its computation in accepting state if and only if the machine in basic model does the same.

This proves the equivalence of both machines with regard to the accepted language. Therefore, Turing machines in basic model are equivalent to Turing machines with the halting accepting state. □

Definition 5.11. Turing machine with halting states is a system

$$M = (Q, \Sigma, \Gamma, \delta, q_0, B, F, R)$$

such that it terminates its computation for any input data, where:
- $F = \{q_A\}$ – includes only one accepting state q_A,
- $R = \{q_R\}$ – includes a special non-accepting state q_R,
- computation for any input always reaches one of states q_A or q_R,
- computation stops if and only if it reaches q_A or q_R,
- other components of the system are as described in Definition 5.1.

Proposition 5.2. *Turing machines with halting states are equivalent to Turing machines in basic model with the stop property.*

Proof. Modify the proof of Proposition 5.1 to justify this Proposition. □

5.1.4 Guarding the tape beginning

Basic model of Turing machines raises a practical problem of how to detect the tape beginning. The machine of Example 5.1 stores the special symbol X in the first cell and finally – when the head is passed to the beginning of the tape – this symbol is turned to output 0. Of course, a general solution of this problem is similar: just to store special symbols in the first cell of the tape, which replaces symbols defined by the transition function. However, such a solution enlarges the set of states, the tape alphabet and the transition function. Model of the Turing machine with guard avoids this problem.

Definition 5.12. Turing machine in basic model with guard is a system

$$M = (Q, \Sigma, \Gamma, \delta, q_0, B, \#, F, C)$$

where:
- $\#$ – the guard symbol (of tape alphabet), $\# \in \Gamma$, $\# \notin \Sigma$,
- other components are as in Definition 5.1,
- the input configuration is shown in Figure 5.2,
- the head can visit the first cell (with guard) but cannot change its contents; that is, must print the guard when is doing transition and shift of the head must be done to the right,
- the head cannot print the guard anywhere besides the first cell.

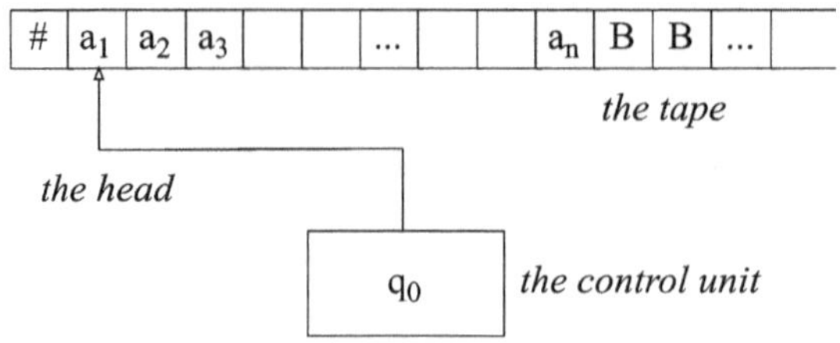

Figure 5.2: Turing machine with guard.

Proposition 5.3. *Turing machines with guard are equivalent to Turing machines in basic model.*

Proof. Given a Turing machine in basic model

$$M = (Q, \Sigma, \Gamma, \delta, q_0, B, F, C)$$

we get its formal equivalent Turing machine with guard by supplementing the system with the guard symbol, shifting input data one cell right, storing the guard symbol in the leftmost cell and placing the head over the second leftmost cell (over the leftmost input symbol). Therefore, we get the following system

$$M_G = (Q, \Sigma, \Gamma \cup \{\#\}, \delta, q_0, B, \#, F, C)$$

Other components of the M_G stay unchanged comparing to M. Any computation of M_G will be precisely the same as for M and the head will never visit the guard cell.

Consider an opposite situation. Given a Turing machine with guard

$$M_G = (Q, \Sigma, \Gamma, \delta, q_0, B, \#, F, C)$$

the following machine is equivalent

$$M' = (Q', \Sigma, \Gamma, \delta', q_0', B, F', C')$$

The machine M' will realize the following computation:
- shifts input data one cell right, stores the guard symbol # in the first cell and leaves the head at the guard symbol (on the leftmost cell),
- simulates computation of M_G,
- shifts the output data one cell left (this deletes the guard symbol in the leftmost cell) and leaves the head at the leftmost output symbol (at the leftmost cell).

A detailed description of M' is given in Example 5.3. □

Example 5.3. Design a Turing machine in basic model equivalent to a given Turing machine with guard.

Solution. Let us assume that a Turing machine with guard is given as

$$M_G = (Q, \Sigma, \Gamma, \delta, q_0, B, \#, F, C)$$

The following Turing machine in basic model is equivalent to M_G

$$M' = (Q', \Sigma, \Gamma, \delta', q_0', B, F', C')$$

where:

- $Q' = Q \cup \{p_0, p_L\} \cup \{p_a : a \in \Sigma\} \cup \{r_0, r', r_A, r_R\} \cup \{r_X : X \in \Gamma - \{\#\}\}$,
- $q_0' = p_0$,
- $F' = \{r_A\}$,
- C' – computation terminates if and only if any of state r_A, r_R is reached.

The part of the transition function, which shifts input data one cell right and inserts the guard symbol, is given below:

- $\delta'(p_0, B) = (q_0, \#, R)$ – when input is empty, just store the guard symbol and shift the head right,
- $\delta'(p_0, a) = (p_a, \#, R)$ for $a \in \Sigma$ – insert the guard symbol, remember an input symbol in a respective state, shift the head right,
- $\delta'(p_a, b) = (p_b, a, R)$ for $a, b \in \Sigma$ – remember an input symbol in a respective state and simultaneously store in the cell the symbol remembered in the state of the previous transition, shift the head right,
- $\delta'(p_a, B) = (p_L, a, L)$ for $a \in \Sigma$ – store in the cell the last input symbol remembered in the state of the previous transition, shift the head left,
- $\delta'(p_L, a) = (p_L, a, L)$ for $a \in \Sigma$ – pass the head to the beginning of the tape,
- $\delta'(p_L, \#) = (q_0, \#, R)$ – shift the head to the first input symbol.

The part of the transition function, which follows the computation of M_G is:

$$\delta'(q, X) = \delta(q, X) \quad \text{for } q \in Q - \{q_A\}, \quad X \in \Gamma$$

The part of the transition function, which shifts output data one cell left and removes the guard symbol, is as follows:

- $\delta'(q_A, B) = (r_B, B, L)$ – output data is empty,
- $\delta'(q_A, X) = (r_0, X, R)$ for $X \in \Gamma - \{B\}$ and $\delta'(r_0, X) = (r_0, X, R)$ for $X \in \Gamma - \{B\}$ – pass the head to the end of output data,
- $\delta'(r_0, B) = (r_B, B, L)$ – remember the blank symbol (the first one right of output data, the leftmost one at the tape) in states, begin passing the head left and shifting output data left,
- $\delta'(r_X, Y) = (r_Y, X, L)$ for $X, Y \in \Gamma - \{\#\}$ – remember an output symbol in a respective state and simultaneously store in the cell the symbol remembered in a state of the previous transition, shift the head left,

- $\delta'(r_X, \#) = (r', X, R)$ for $X \in \Gamma - \{\#\}$ – remove the guard symbol, store the leftmost output symbol instead, shift the head right (this is the leftmost cell, no shift left)
- $\delta'(r', X) = (r_A, X, L)$ for $X \in \Gamma - \{\#\}$ – return to the leftmost cell.

The transition function is undefined for any arguments not considered above. The machine is assumed to fall into infinite computation if not accepts an input. A discussion on the adaptation of this solution to the transition function undefined for some arguments is left to the reader.

Remark 5.3. The other models of Turing machines, e. g. Turing machines with the stop property or Turing machines with halting states, can be turned to models with guard. Formulation and proof of equivalence of models with guard and other models of Turing machines are analogous to the proof of Proposition 5.3.

5.1.5 Turing machines with a multi-track tape

A Turing machine with multi-track tape has the tape split at its length to a given number of tracks. Every track is split into cells and is one way infinite. The head reads symbols of all cells of the same slice (column) and shifts simultaneously over all tracks. An initial configuration of a Turing machine k-tracks tape is shown in Figure 5.3. Input data is stored in the beginning cells of track number one while all other cells of track number one and all cells of other tracks are filled in with the blank symbol B.

Definition 5.13. Turing machine in basic model with a multi-track (k-tracks) tape is a system

$$M = (Q, \Sigma, \Gamma, \delta, q_0, B, F, C)$$

where:
- $\delta : Q \times \Gamma^k \to Q \times \Gamma^k \times \{L, R\}$ is the transition function,
- input data $w = a_1 a_2 \ldots a_n$ is represented as sequence of k-tuples $((a_1, B, \ldots, B), (a_2, B, \ldots, B), \ldots, (a_n, B, \ldots, B))$, every k-tuple fills in one slice of the tape, cf. Figure 5.3,
- output data is represented in a way similar to the representation of input data, that is, it is stored in the beginning cells of the first track while all other cells are filled in with the blank symbol,
- other components are as in Definition 5.1.

Proposition 5.4. *Turing machines with a multi-track tape are equivalent to Turing machines in basic model.*

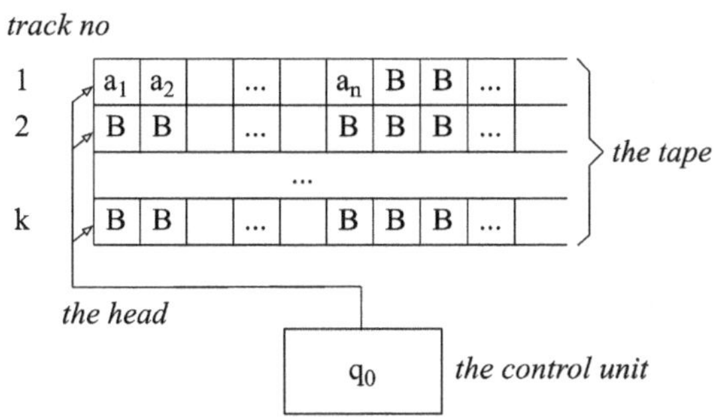

Figure 5.3: Turing machine in basic model with a multi-track tape.

Proof. First of all, a Turing machine in basic model is a special case of a Turing machine with a multi-track tape having just one track.

Secondly, a Turing machine with k-tracks tape:

$$M' = (Q, \Sigma', \Gamma', \delta, q_0, B', F, C)$$

is equivalent to the following Turing machine in basic model:

$$M = (Q, \Sigma, \Gamma, \delta, q_0, B, F, C)$$

where:
- $\Sigma = \Sigma' \times \{B\} \times \{B\} \times \cdots \times \{B\}$ – product of k sets
- $\Gamma = \Gamma' \times \Gamma' \times \cdots \times \Gamma'$ product of k sets
- $B = (B', B', \ldots, B')$ – k-tuple
- other components are as in machine M'. □

The above conclusion is not surprising. Turing machines with multi-track tape work in a way quite similar to Turing machines in basic model, that is, they have one-way infinite tape, their one head reads and writes data of the whole slice at a time, etc. Therefore, when data of a slice is interpreted as one symbol of a tape alphabet, a Turing machine with a multi-track tape is just a Turing machine in basic model. Turing machines with multi-track tape are essentially useful in proofs of equivalence of other, more important, models of Turing machines. This is the main motivation for discussing this model.

5.1.6 Turing machines with two-way infinite tape

Variations of Turing machines discussed so far are slightly modified Turing machines in basic model. Two-way infinite tape is the first important modification of Turing machines in basic model, cf. Figure 5.4. When Remark 5.2 is employed, Turing machines with two-way infinite tape permit avoiding problems with passing the head to the beginning of the tape or to the beginning of data stored on the tape: the first cell with the

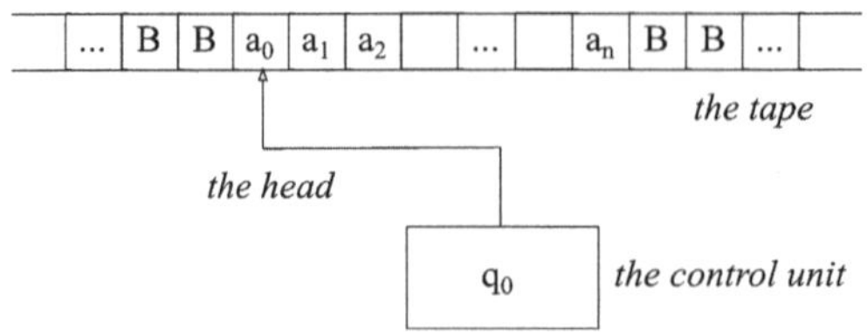

Figure 5.4: Turing machine with two-way infinite tape.

blank symbol prior to nonblank symbols indicates the beginning of data stored on the tape (of course, assuming that blank symbol cannot appear between nonblank ones).

Definition of Turing machine with two-way infinite tape is identical with Definition 5.1. Turing machine with two-way infinite tape does not need to identify the beginning of the tape. However, the definition of configuration (step description) of this type of machines is slightly different from that of machines in basic model. The difference is in the description of the left sequence of symbols.

Definition 5.14. A step description (a configuration) of a Turing machine with two-way infinite tape $M = (Q, \Gamma, \Sigma, \delta, q_0, B, F, C)$ is the following sequence of symbols:

$$\alpha_1 \; q \; \alpha_2$$

where:
- $q \in Q$ and α_2 are the same as in Definition 5.2,
- α_1 is the sequence of symbols stored in cells left of the head, starting with a leftmost non-blank symbol and ending with the symbol in the cell before the one under the head.

Proposition 5.5. *Turing machines with two-way infinite tape are equivalent to Turing machines in basic model.*

Proof. We prove that Turing machines with two-way infinite tape and Turing machines with multi-track tape are equivalent. Because Turing machines with a multi-track tape are equivalent to Turing machines in basic model, cf. Proposition 5.13, then we get equivalence declared in this Proposition.

For a given Turing machine in basic model M_1, an equivalent Turing machine with two-way infinite tape M_2 is equal to M_1. Computation of M_1 is being done on the *right half* of tape (the right part of the tape which begins with the cell holding the first symbol of input data).

Assuming that a machine with two-way infinite tape is given as follows:

$$M_2 = (Q_2, \Sigma, \Gamma_2, \delta_2, q_0^2, B_2, F, C)$$

we will design a machine with a multi-track tape:

$$M_1 = (Q_1, \Sigma, \Gamma_1, \delta_1, q_0^1, B_1, ¢, F, C)$$

First of all, a method of representation of a two-way infinite tape must be found out. A one-way infinite tape is a 2-tracks tape. The *right half* of the two-way infinite tape matches the upper track and the *left half* rotated by 180° matches the lower track, cf. Figure 5.4 and Figure 5.5

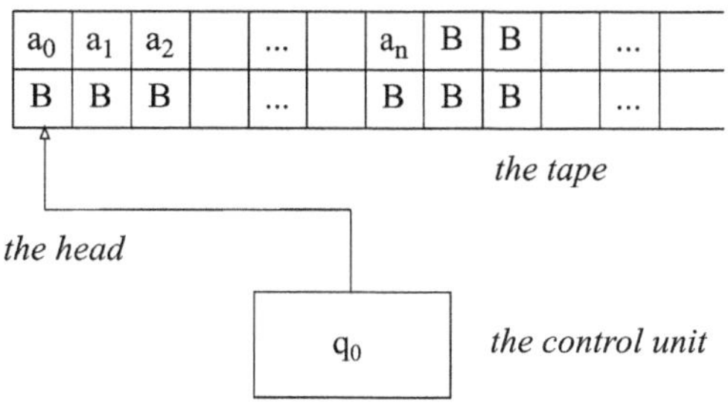

Figure 5.5: Turing machine with two-track tape – simulation of two-way infinite tape.

Computation of M_2 done on the *right half* of tape is followed on the upper track of M_1. Computation of M_2 done on the *left half* of tape is followed on the lower track of M_1 with the head shifting oppositely than the head of M_2. Note that the first cell of the lower track plays the role of the guard while the content of the *left half* of tape is stored beginning with the second cell, cf. Figure 5.6.

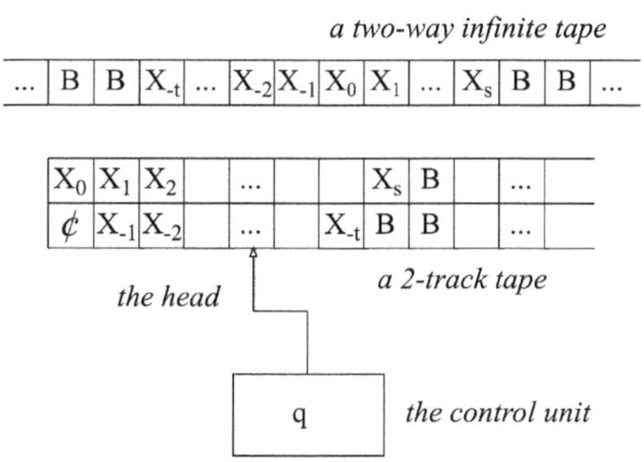

Figure 5.6: Turing machine with two-track tape – simulation of computation of Turing machine with two-way infinite tape.

A formal and detailed description of M_1 simulating M_2 is as follows:

- $Q_1 = Q_2 \times \{U, B\} \cup \{q_0^1\}$
- $\Gamma_1 = \Gamma_2 \times \Gamma_2 \cup \Gamma_2 \times \{¢\}$
- $\Sigma_1 = \Sigma_2 \times \{B_2\}$
- $F_1 = \{(q, U), (q, L) : q \in F_2\}$
- $B_1 = (B_2, B_2)$

Symbols U and L depict placement of the head of M_2 on the *right* and the *left half* of the tape.

The first transition of M_1 stores the guard symbol in the first cell of the lower track and simulates the first transition of M_2 going to either the upper or the lower track:

$$\delta_1(q_0^1, (a_0, B)) = \begin{cases} ((p,U),(Y,¢),R) & \text{if } \delta_2(q_0^2, a_0) = (p,Y,R) \\ ((p,L),(Y,¢),R) & \text{if } \delta_2(q_0^2, a_0) = (p,Y,L) \end{cases}$$

When the head of M_1 is placed right of the first cell (computation of M_2 cannot change current *half* of the tape), then:

$$\left.\begin{aligned} \delta_1((q,U),(X,Z)) &= ((p,U),(Y,Z),D) \\ \delta_1((q,L),(Z,X)) &= ((p,L),(Z,Y),\overline{D}) \end{aligned}\right\} \quad \text{if } \delta_2\,(q,X) = (p,Y,D)$$

where: D is a direction of the head shift, $\overline{D}$ is opposite to D.

When the head of M_1 is placed at the first cell (computation of M_2 may change current *half* of the tape), then:

$$\left.\begin{aligned} \delta_1((q,U),(X,¢)) &= ((p,U),(Y,¢),R) \\ \delta_1((q,L),(X,¢)) &= ((p,U),(Y,¢),R) \end{aligned}\right\} \quad \text{if } \delta_2\,(q,X) = (p,Y,R)$$

$$\left.\begin{aligned} \delta_1((q,U),(X,¢)) &= ((p,L),(Y,¢),R) \\ \delta_1((q,L),(X,¢)) &= ((p,L),(Y,¢),R) \end{aligned}\right\} \quad \text{if } \delta_2\,(q,X) = (p,Y,L)$$

where,

- U and L, coming in states labels, stand for *upper* and *lower* track,
- L and R, coming as the third element of the transition function (3-tuple) value, stand for *left* and *right* direction of head shifts. □

In light of Proposition 5.4 and Proposition 5.5 the following conclusion is fairly obvious:

Proposition 5.6. *Turing machines with a multi-track two-way infinite tape are equivalent to Turing machines in basic model.*

5.1.7 Multi-tape Turing machines

A multi-tape Turing machine, cf. Figure 5.7, satisfies the assumptions:

- it has several tapes and one head for each tape,
- tapes are two-way infinite,
- the initial configuration assumes:
  - the control unit is in the initial state,
  - input data is stored on the first tape,
  - the head of the first tape is placed over the first (leftmost) symbol of input data,
  - all other cells of the first tape and all cells of other tapes are filled in with the blank symbol,

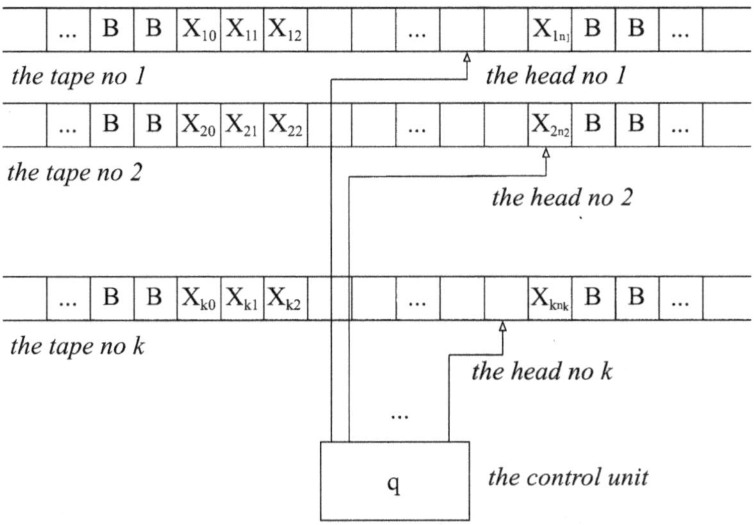

Figure 5.7: A multi-tape Turing machine.

- a transition of a multi-tape Turing machine is as follows:
  - the control unit switches to some state,
  - every head stores a tape symbols in its cell,
  - every head shifts left, right or stays in current position independently on other heads.

Formal definition of a multi-tape Turing machine is as follows:

Definition 5.15. A multi-tape Turing machine (with k-tapes) is a system

$$M = (Q, \Sigma, \Gamma_1 \times \Gamma_2 \times \cdots \times \Gamma_k, \delta, q_0, B, F, C)$$

where:
- $\Gamma_1, \Gamma_2 \ldots, \Gamma_k$ are alphabets of tapes. It is assumed that all tapes have the same alphabet Γ, if not stated differently,
- $\delta : Q \times (\Gamma_1 \times \Gamma_2 \times \cdots \times \Gamma_k) \rightarrow Q \times (\Gamma_1 \times \Gamma_2 \times \cdots \times \Gamma_k) \times \{L, R, S\}^k$ is the transition function with directions of head shift: left, right and stop (no shift of a head),
- other components are as in Definition 5.1.

Proposition 5.7. *Multi-tape Turing machines are equivalent to Turing machines in basic model.*

Proof. A Turing machine in basic model is equivalent to some Turing machine with two-way infinite tape, cf. Proposition 5.5. Moreover, a Turing machine with two-way infinite tape is a case of a multi-tape Turing machine; it is just a multi-tape Turing machine with one tape. Therefore, for any Turing machine in basic model, we can directly find an equivalent multi-tape Turing machine.

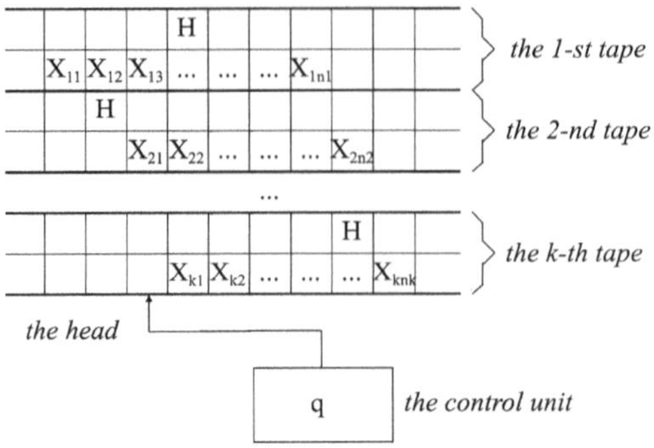

Figure 5.8: Turing machine with a multi-track two-ways infinite tape simulating a multi-tape Turing machine. The blank symbol is not printed for the sake of clarity; that is, all cells which are empty in this Figure hold the blank symbol.

Now, we give an idea of designing a Turing machine with a multi-track two-way infinite tape M_1 for a given k-tapes Turing machine M_m:

- k-tapes are represented on a two-way infinite tape with 2-k-tracks; each tape corresponds to a pair of tracks, cf. Figure 5.8
  - content of every tape is stored on the bottom track of the corresponding pair of tracks,
  - the special symbol H stored in a cell of upper track marks the head position of the tape of the machine M_m,
  - all other cells of both tracks are filled in with the empty symbol B,
- initial configuration of M_1 is as follows:
  - the initial state of M_m corresponds to the relevant state of M_1,
  - input data stored on the first tape of a multi-tape Turing machine is represented on the lower track of the first pair of tracks of M_1,
  - the head markers of all tapes of M_m are placed in the tape slice holding the first symbol of input data,
- a transition of the machine M_m is simulated by several transitions of the machine M_1:
  - the head of M_1 passes from the tape slice holding the leftmost head marker H to the tape slice holding the rightmost head marker. This is so-called *collect-data* pass. Information about symbols under all tape markers (under heads of M_m) is collected during this pass and remembered in a relevant state of M_1,
  - the transition function of M_m is applied to collected data (state of M_m and symbols under heads of M_m). The result of the transition function is remembered in a state of M_1. Recall that transition function of M_m returns: a state, symbols to be printed by heads and directions of heads shifts,
  - the head of M_1 is passed to the time slice holding the leftmost tape marker H. This is so-called *update* pass. Symbols under head markers and head markers'

positions are updated during this pass. Updates are done according to the value of the transition function obtained in the previous point,
- computation of M_1 is terminated if and only if the stop condition of M_m is satisfied. Input data of M_1 is accepted if and only if M_m terminates computation in an accepting state. □

Note that simulation of a multi-tape Turing machine by Turing machines with one tape requires a huge set of states, which makes transition function highly enlarged. It also increases the number of transitions for the same input data.

As it is stated in Proposition 5.7, data collected during the left to right pass of the head on simulating machine are stored in the set of states. So, a normal set of states Q_1 of the simulating machine should be extended by the Cartesian product of:
- a pass direction: $Q_1 \times \{C, U\}$ to indicate if this is *the collect-data* pass or *the update* pass,
- the set of states of a multi-tape machine: $Q_1 \times \{C, U\} \times Q_m$ to keep a state of M_m during both passes,
- alphabets of tapes: $Q_1 \times \{C, U\} \times Q_m \times \Gamma_1 \times \Gamma_2 \times \cdots \times \Gamma_k$ for symbols under heads during both passes,
- markers of heads, which were visited in the collect-data pass, this might be done by extending tapes' alphabets with a special *the marker-not-visited-yet* symbol $\hbar$: $Q_1 \times \{C, U\} \times Q_m \times \Gamma'_1 \times \Gamma'_2 \times \cdots \times \Gamma'_k$, where $\Gamma'_i = \Gamma_i \cup \{\hbar\}$ for any tape i,
- a value of the transition function δ_m is stored in already included components $Q_m \times \Gamma'_1 \times \Gamma'_2 \times \cdots \times \Gamma'_k$ with the update pass indication,
- heads' shifts $Q_1 \times \{C, U\} \times Q_m \times \Gamma'_1 \times \Gamma'_2 \times \cdots \times \Gamma'_k \times \{L, R, S\}^k$,
- update heads' positions counters would require four symbols to indicate status (not updated, being shifted left, being shifted right, updated), $Q_1 \times \{C, U\} \times Q_m \times \Gamma'_1 \times \cdots \times \Gamma'_k \times \{L, R, S\}^k \times \{X^i, X^{ii}, X^{iii}, X^{iv}\}^k$ for each tape,
- and more states is required to organize simulation details.

Thus, states could be labeled by elements of the above Cartesian product, that is, by $(3k+3)$-tuples. So then, the number of states is not less than $r = 2 * 3^k * 3^k * |Q_1| * |Q_m| * |\Gamma'_1| * |\Gamma'_2| * \cdots * |\Gamma'_k|$

On the other hand, symbols of the tape alphabet of a simulating machine could be denoted by all tuples of the Cartesian product of tapes' alphabets and symbols stored in tracks with heads' markers (the blank symbol B and the head's marker symbol H) $\{H, B\} \times \Gamma_1 \times \{H, B\} \times \Gamma_2 \times \cdots \times \{H, B\} \times \Gamma_k$. Thus, the number of tape symbols of simulating machine is equal to $c = 2^k * |\Gamma_1| * |\Gamma_2| * \cdots * |\Gamma_k|$.

Finally, the size of the transition table of the simulating machine is not less than $r * c$ comparing to the size of the transition table of a multi-tape Turing machine, which is equal to $|Q_m| * (|\Gamma_1| * |\Gamma_2| * \cdots * |\Gamma_k|)$.

Let us analyze the possible increase in the number of transitions. The worse case is when the head of one tape of a multi-tape Turing machine always shifts right and the

head of another tape always shifts left. Tape slices with these two heads' markers will be separated by:
- one tape slices after simulation of the first transition,
- three slices after simulation of two transitions,
- $2 * n - 1$ tape slices after simulation of n transitions.

Simulation of transitions of a multi-tape Turing machine, as described above, requires the following numbers of transitions of simulating machine:
- three transitions of M_1 to simulate the first transition of M_m,
- at least seven transitions of M_1 to simulate the second transition of M_m,
- ...
- at least $4 * n - 1$ transitions of M_1 to simulate the n-th transition of M_m. This number will be increased by transitions updating markers of those heads of a multi-tape machine, which shift right. The increment will not exceed $2 * (k - 2)$, where k is the number of tapes.

As a result, simulation of n transitions of a multi-tape Turing machine requires not less than

$$\begin{aligned} f(n) &= 3 + 7 + 11 + \cdots + (4 * n - 1) + n * (k - 2) \\ &= 4 + 8 + \cdots + 4 * n + n * (k - 3) = 2 * n^2 + (k - 1) * n \end{aligned}$$

transitions of simulating machine. This number could be estimated: $2 * n^2 \le f(n) \le 3 * n^2$ for n big enough. Therefore, we can say that simulation increases cost of computation with square.

Configuration of a multi-tape Turing machine

Definition of a configuration (step description) of a multi-tape machine must consider contents of several tapes and one state of control unit:

Definition 5.16. A configuration (step description) of a multi-tape Turing machine with k-tapes is the following sequence of symbols

$$(\alpha_1^1, \alpha_1^2, \ldots, \alpha_1^k)\ q\ (\alpha_2^1, \alpha_2^2, \ldots, \alpha_2^k)$$

where:
- q is a state of control unit,
- α_1^i is a sequence of symbols for i-th tape in cells preceding the head (like in Turing machines with two-way infinite tape),
- α_2^i is a sequence of symbols for i-th tape in cells from the head's right (like in Turing machines with two-way infinite tape).

In practice, step description will be shown in two ways presented below in examples:

- state of the control unit plays a role of tape markers, contents of tapes is split for left and right sequences of symbols and contents of tapes may be shifted independently each from others:

$$\begin{array}{rcl} & & X_4^1X_5^1X_6^1X_7^1 \\ X_1^1X_2^1X_3^1 & q & X_6^2X_7^2 \\ X_1^2X_2^2X_3^2X_4^2X_5^2 & & BBX_1^3X_2^3X_3^3X_4^3 \end{array}$$

- state of the control unit is placed at the beginning, contents of tapes are fixed together, heads' positions are underscored:

$$\begin{array}{cl} & X_1^1X_2^1X_3^1\underline{X_4^1}X_5^1X_6^1X_7^1 \\ q & X_1^2X_2^2X_3^2X_4^2X_5^2\underline{X_6^2}X_7^2 \\ & \underline{B}BX_1^3X_2^3X_3^3X_4^3 \end{array}$$

Example 5.4. Design a Turing machine computing sum of two binary numbers.

Solution. Let us build a 2-tape Turing machine. According to the general assumption, input data, two nonnegative binary numbers separated with the blank symbol, is stored on the first tape (unlike it is assumed in Remark 5.2, for the sake of limiting the size of the transition function, we use the blank symbol to separate input numbers instead of extra tape symbol). The head of the first tape is placed over the cell holding the first (most significant) digit of the first number. The following algorithm is applied:
- the first number is moved to the second tape,
- heads of both tapes are placed over the last (least significant) digits of both numbers,
- addition is done for consecutive digits of both numbers from the least significant to the most significant digits:
  - digits read by heads and carry are added, the result is stored by the head of the first tape, a carry is remembered in states,
  - the result of addition is stored by the head of the first tape, the head of the second tape stores the blank symbol; then both heads shift left,
  - the process of adding is terminated as soon as both heads read the blank symbol and there is no carry.

A 2-tape Turing machine with halting accepting state implementing this algorithm is as follows:

$$M = (\{q_0, q_1, q_2^0, q_2^1, q_A\}, \{0,1\}, \{0,1,B\}, \delta, q_0, B, \{q_A\})$$

Both tapes have the same alphabet $\Gamma = \{0,1,B\}$. The transition function is given in Table 5.3

A detailed description of the machine is given in the form of comments to the purpose of states:

Table 5.3: The transition function of the Turing machine considered in Example 5.4.

δ	$\begin{matrix} 0 \\ B \end{matrix}$	$\begin{matrix} 1 \\ B \end{matrix}$	$\begin{matrix} B \\ 0 \end{matrix}$	$\begin{matrix} B \\ 1 \end{matrix}$	$\begin{matrix} B \\ B \end{matrix}$
q_0	$\left(q_0, \begin{matrix} B & R \\ 0, & R \end{matrix}\right)$	$\left(q_0, \begin{matrix} B & R \\ 1, & R \end{matrix}\right)$			$\left(q_1, \begin{matrix} B & R \\ B, & S \end{matrix}\right)$
q_1	$\left(q_1, \begin{matrix} 0 & R \\ B, & S \end{matrix}\right)$	$\left(q_1, \begin{matrix} 1 & R \\ B, & S \end{matrix}\right)$			$\left(q_2^0, \begin{matrix} B & L \\ B, & L \end{matrix}\right)$
q_2^0	$\left(q_2^0, \begin{matrix} 0 & L \\ B, & L \end{matrix}\right)$	$\left(q_2^0, \begin{matrix} 1 & L \\ B, & L \end{matrix}\right)$	$\left(q_2^0, \begin{matrix} 0 & L \\ B, & L \end{matrix}\right)$	$\left(q_2^0, \begin{matrix} 1 & L \\ B, & L \end{matrix}\right)$	$\left(q_A, \begin{matrix} B & R \\ B, & S \end{matrix}\right)$
q_2^1	$\left(q_2^0, \begin{matrix} 1 & L \\ B, & S \end{matrix}\right)$	$\left(q_2^1, \begin{matrix} 0 & L \\ B, & S \end{matrix}\right)$	$\left(q_2^0, \begin{matrix} 1 & L \\ B, & L \end{matrix}\right)$	$\left(q_2^1, \begin{matrix} 0 & L \\ B, & L \end{matrix}\right)$	$\left(q_2^0, \begin{matrix} 1 & L \\ B, & S \end{matrix}\right)$

δ	$\begin{matrix} 0 \\ 0 \end{matrix}$	$\begin{matrix} 0 \\ 1 \end{matrix}$	$\begin{matrix} 1 \\ 0 \end{matrix}$	$\begin{matrix} 1 \\ 1 \end{matrix}$
q_0				
q_1				
q_2^0	$\left(q_2^0, \begin{matrix} 0 & L \\ B, & L \end{matrix}\right)$	$\left(q_2^0, \begin{matrix} 1 & L \\ B, & L \end{matrix}\right)$	$\left(q_2^0, \begin{matrix} 1 & L \\ B, & L \end{matrix}\right)$	$\left(q_2^1, \begin{matrix} 0 & L \\ B, & L \end{matrix}\right)$
q_2^1	$\left(q_2^0, \begin{matrix} 1 & L \\ B, & L \end{matrix}\right)$	$\left(q_2^1, \begin{matrix} 0 & L \\ B, & L \end{matrix}\right)$	$\left(q_2^1, \begin{matrix} 0 & L \\ B, & L \end{matrix}\right)$	$\left(q_2^1, \begin{matrix} 1 & L \\ B, & L \end{matrix}\right)$

- q_0 – the first number is moved to the second tape,
- q_1 – the head of the first tape is passed to the end of data, heads of both tapes are placed at the least significant digits of numbers stored on tapes,
- q_2^0, q_2^1 – superscripts are used to remember carry, adding both numbers, heads of both tapes are simultaneously passed left, corresponding digits of numbers and the carry are added step by step.

Note that the Turing machine designed above can be formally turned to a one with the stop condition. It can be done by filling empty entries of Table 5.3 with a transition to the halting rejecting state.

Finally, let us simulate computation for given numbers 1101 and 101:

$$q_0^{1101B101} \succ {}_{1}q_0^{101B101} \succ {}_{11}q_0^{01B101} \succ {}_{110}q_0^{1B101}$$

$$\succ {}_{1101}q_0^{B101} \succ {}_{1101}q_1^{101} \succ {}^{1}_{1101}q_1^{01} \succ {}^{10}_{1101}q_1^{1} \succ {}^{101}_{1101}q_1$$

$$\succ {}^{10}_{110}{q_2^0}^{1}_{1} \succ {}^{1}_{11}{q_2^1}^{00}_{0} \succ {}_{1}{q_2^0}^{110}_{1} \succ {q_2^1}^{B010}_{1} \succ {q_2^1}^{B0010}$$

$$\succ {q_2^0}^{B10010} \succ q_A^{10010}$$

5.1.8 Programming with Turing machines

Implementation of algorithms in the form of computer programs employs such methods as modularity or top-down programming. Programming techniques use collaborating units such as procedures or functions. As mentioned before, Turing machines can serve as tools for solving problems. They can be seen as computational tools, with transition function being a low-level programming language. Indeed, Turing machines can be organized as collaborating units in solving problems. Remark 5.1 allows for easy data passing from one machine to another.

An example of linking Turing machines is given in Example 5.3. The solution is an example of three collaborating Turing machines: a machine shifting input data right and inserting the guard symbol, a given machine with guard, a machine shifting output data right and removing the guard symbol.

Example 5.5. Design a Turing machine computing product of two natural numbers.

Solution. We assume that factors are positive numbers. This assumption permits avoiding some details and not increase the number of states.

The product of two natural numbers will be computed as multiple additions of one factor, with another factor being a counter of additions. The solution is based on three Turing machines, which are finally joined into the final solution. The designed machines are two-tape machines with the same input alphabet. The final Turing machine has two tapes and one transition table with three groups of states, one group for each component machine. However, the below transition table of the joined Turing machine stays split into three separated parts.

The first Turing machine

$$M_P = (Q_P = \{p_1, p_2, p_L, p_L^2, r_0\},\ \{0,1\},\ \{0,1,B\},\ \delta_P, p_1, B, \{r_0\})$$

with transition table given in Table 5.4, prepares input data for further processing. Input data consists of two positive binary numbers separated with the blank symbol (what breaks assumptions of Remark 5.2 that blank symbol cannot be placed between nonblank ones, but significantly reduces the size of transition tables of designed component machines). This machine makes the copy of input data on the second tape, erases the all but last symbols of the second input argument, that is, replaces these symbols with the blank symbol. The last symbol of the second input argument (rightmost nonempty symbol) is replaced with the digit 1. An input configuration of tapes is shown in Figure 5.9 a). An output configuration of this machine is shown in Figure 5.9 b).

A few comments on states' purpose to clarify the design:

- p_1 – copy the first argument (the first factor of the computed product) to the second tape, shifting both heads right, the original first argument (on the first tape) is an initial value of the result,
- p_2 – move the second argument (the second factor of the product) to the second tape, shifting both heads right,

Table 5.4: The transition function of the Turing machine M_R preparing input data for further processing.

δ_P	$\begin{matrix}0\\B\end{matrix}$	$\begin{matrix}1\\B\end{matrix}$	$\begin{matrix}B\\0\end{matrix}$	$\begin{matrix}B\\1\end{matrix}$	$\begin{matrix}B\\B\end{matrix}$	$\begin{matrix}0\\0\end{matrix}$	$\begin{matrix}0\\1\end{matrix}$	$\begin{matrix}1\\0\end{matrix}$	$\begin{matrix}1\\1\end{matrix}$
p_1	$\left(p_1, \begin{matrix}0\ R\\ 0, R\end{matrix}\right)$	$\left(p_1, \begin{matrix}1\ R\\ 1, R\end{matrix}\right)$			$\left(p_2, \begin{matrix}B\ R\\ B, R\end{matrix}\right)$				
p_2	$\left(p_2, \begin{matrix}B\ R\\ 0, R\end{matrix}\right)$	$\left(p_2, \begin{matrix}B\ R\\ 1, R\end{matrix}\right)$			$\left(p_L, \begin{matrix}B\ L\\ B, L\end{matrix}\right)$				
p_L			$\left(p_L^2, \begin{matrix}1\ L\\ 0, L\end{matrix}\right)$	$\left(p_L^2, \begin{matrix}1\ L\\ 1, L\end{matrix}\right)$					
p_L^2			$\left(p_L^2, \begin{matrix}B\ L\\ 0, L\end{matrix}\right)$	$\left(p_L^2, \begin{matrix}B\ L\\ 1, L\end{matrix}\right)$	$\left(r_0, \begin{matrix}B\ S\\ B, S\end{matrix}\right)$				

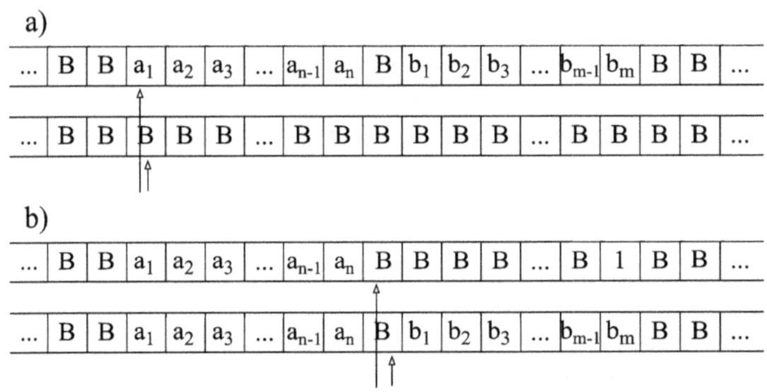

Figure 5.9: Tapes' configurations of Turing machines built in Example 5.5.

- p_L – initialize the additions' counter at the first tape with 1, that is, 1 is stored in the cell of the rightmost not blank input symbol (it keeps compatibility with the initial value of the result),
- q_L^2 – shift both heads Left to place them between factors,
- r_0 – terminate computation and accept output data (when machines are joined, this is initial state of the second Turing machine).

The second Turing machine

$$M_R = (Q_R = \{r_0, r_P, r_P^1, q_L^1, q_L^2, r_D^1, r_D^2, r_A, q_0\}, \{0,1\}, \{0,1,B\}, \delta_R, r_0, B, \{r_A\})$$

checks if the final result is computed, that is, if the additions' counter on the first tape is equal to the second factor stored on the second tape. If not, then the additions' counter is incremented by one and computation is passed to the third machine in order to add the first factor to the final result. The transition table of this machine is provided in Table 5.5

Table 5.5: The transition function of the Turing machine M_R checking whether the final result is computed.

δ	$\begin{matrix}0\\B\end{matrix}$	$\begin{matrix}1\\B\end{matrix}$	$\begin{matrix}B\\0\end{matrix}$	$\begin{matrix}B\\1\end{matrix}$	$\begin{matrix}B\\B\end{matrix}$
r_0					$\left(r_P, \begin{matrix}B & R\\ B' & R\end{matrix}\right)$
r_P			$\left(r_P^1, \begin{matrix}B & R\\ 0' & R\end{matrix}\right)$	$\left(r_P^1, \begin{matrix}B & R\\ 1' & R\end{matrix}\right)$	$\left(r_L^2, \begin{matrix}B & L\\ B' & L\end{matrix}\right)$
r_P^1			$\left(r_P^1, \begin{matrix}B & R\\ 0' & R\end{matrix}\right)$	$\left(r_P^1, \begin{matrix}B & R\\ 1' & R\end{matrix}\right)$	$\left(r_D^1, \begin{matrix}B & L\\ B' & L\end{matrix}\right)$
r_L^2					$\left(r_L^1, \begin{matrix}B & L\\ B' & L\end{matrix}\right)$
r_L^1	$\left(r_L^1, \begin{matrix}0 & L\\ B' & L\end{matrix}\right)$	$\left(r_L^1, \begin{matrix}1 & L\\ B' & L\end{matrix}\right)$			$\left(r_A, \begin{matrix}B & R\\ B' & S\end{matrix}\right)$
r_D^1			$\left(r_D^2, \begin{matrix}1 & L\\ 0' & L\end{matrix}\right)$	$\left(r_D^2, \begin{matrix}1 & L\\ 1' & L\end{matrix}\right)$	$\left(q_0, \begin{matrix}B & L\\ B' & L\end{matrix}\right)$
r_D^2			$\left(r_D^2, \begin{matrix}B & L\\ 0' & L\end{matrix}\right)$	$\left(r_D^2, \begin{matrix}B & L\\ 1' & L\end{matrix}\right)$	$\left(q_0, \begin{matrix}B & L\\ B' & L\end{matrix}\right)$

δ_P	$\begin{matrix}0\\0\end{matrix}$	$\begin{matrix}0\\1\end{matrix}$	$\begin{matrix}1\\0\end{matrix}$	$\begin{matrix}1\\1\end{matrix}$
r_0				
r_P	$\left(r_P, \begin{matrix}0 & R\\ 0' & R\end{matrix}\right)$	$\left(r_P^1, \begin{matrix}0 & R\\ 1' & R\end{matrix}\right)$	$\left(r_P^1, \begin{matrix}1 & R\\ 0' & R\end{matrix}\right)$	$\left(r_P, \begin{matrix}1 & R\\ 1' & R\end{matrix}\right)$
r_P^1	$\left(r_P^1, \begin{matrix}0 & R\\ 0' & R\end{matrix}\right)$	$\left(r_P^1, \begin{matrix}0 & R\\ 1' & R\end{matrix}\right)$	$\left(r_P^1, \begin{matrix}1 & R\\ 0' & R\end{matrix}\right)$	$\left(r_P^1, \begin{matrix}1 & R\\ 1' & R\end{matrix}\right)$
r_L^2	$\left(r_L^2, \begin{matrix}B & L\\ B' & L\end{matrix}\right)$			$\left(r_L^2, \begin{matrix}B & L\\ B' & L\end{matrix}\right)$
r_L^1	$\left(r_L^1, \begin{matrix}0 & L\\ B' & L\end{matrix}\right)$	$\left(r_L^1, \begin{matrix}0 & L\\ B' & L\end{matrix}\right)$	$\left(r_L^1, \begin{matrix}1 & L\\ B' & L\end{matrix}\right)$	$\left(r_L^1, \begin{matrix}1 & L\\ B' & L\end{matrix}\right)$
r_D^1	$\left(r_D^2, \begin{matrix}1 & L\\ 0' & L\end{matrix}\right)$	$\left(r_D^2, \begin{matrix}1 & L\\ 1' & L\end{matrix}\right)$	$\left(r_D^1, \begin{matrix}1 & L\\ 0' & L\end{matrix}\right)$	$\left(r_D^1, \begin{matrix}1 & L\\ 1' & L\end{matrix}\right)$
r_D^2	$\left(r_D^2, \begin{matrix}0 & L\\ 0' & L\end{matrix}\right)$	$\left(r_D^2, \begin{matrix}0 & L\\ 1' & L\end{matrix}\right)$	$\left(r_D^2, \begin{matrix}1 & L\\ 0' & L\end{matrix}\right)$	$\left(r_D^2, \begin{matrix}1 & L\\ 1' & L\end{matrix}\right)$

Input data is obtained from the first machine for the first time; then, it is obtained from the third machine in order to check the number of additions. Output data is passed to the third machine to add the first factor again if it contributes to final results less than the second factor. Otherwise, the product has been computed. An input configuration of tapes of this machine is shown in Figure 5.9 b) and Figure 5.10 b). An output configu-

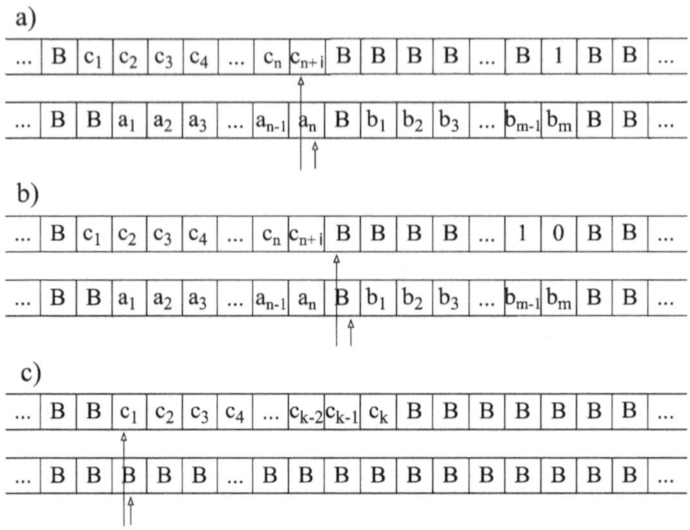

Figure 5.10: Tapes' configurations of Turing machines built in Example 5.5.

ration of this machine is shown in Figure 5.10 a). The final configuration is shown in Figure 5.10 c).

A detailed description of the machine is given in the form of comments to the purpose of states:

- r_0 – initiate computation, shift heads right, switch to the state r_P in order to begin comparing the current value of additions' counter with the second factor,
- r_P – move heads right, com*P*are the current value of additions' counter with the second factor,
- r_L^2 – comparing in the state r_P shows that the current value of additions' counter and the second factor are equal. Therefore, move heads *L*eft, clear the addition counter and the *2*nd factor,
- r_L^1 – continue passing heads *L*eft, clear the *1*st factor on the second tape, place the head of the first tape on the leftmost output symbol,
- r_P^1 – com*P*aring shows that the current value of additions' counter is less than the second factor, continue passing heads right, in order to increment the counter by *1*,
- r_D^1 – a*D*ding *1* to the additions' counter, heads passes left,
- r_D^2 – continue passing heads left,
- q_0 – pass control to the machine computing sum of two numbers.

The third Turing machine

$$M_Q = (Q_Q\{q_0, q_1, q_R, r_0\}, \{0,1\}, \{0,1,B\}, \delta_Q, q_0, B, \{r_0\})$$

with transition table given in Table 5.6 adds the first factor (kept on the second tape) increasing the result on the first tape. Input data of this machine consists of the tempo-

Table 5.6: The transition function of the Turing machine adding the first factor to the current result.

δ	$\begin{matrix}0\\B\end{matrix}$	$\begin{matrix}1\\B\end{matrix}$	$\begin{matrix}B\\0\end{matrix}$	$\begin{matrix}B\\1\end{matrix}$	$\begin{matrix}B\\B\end{matrix}$
q_0	$\left(q_0, \begin{matrix}0 & L\\ B, & L\end{matrix}\right)$	$\left(q_0, \begin{matrix}1 & L\\ B, & L\end{matrix}\right)$			$\left(q_R, \begin{matrix}B & R\\ B, & R\end{matrix}\right)$
q_1	$\left(q_0, \begin{matrix}1 & L\\ B, & S\end{matrix}\right)$	$\left(q_1, \begin{matrix}0 & L\\ B, & S\end{matrix}\right)$			$\left(q_0, \begin{matrix}1 & L\\ B, & S\end{matrix}\right)$
q_R	$\left(q_R, \begin{matrix}0 & R\\ B, & R\end{matrix}\right)$	$\left(q_R, \begin{matrix}1 & R\\ B, & R\end{matrix}\right)$			$\left(r_0, \begin{matrix}B & S\\ B, & S\end{matrix}\right)$

δ_P	$\begin{matrix}0\\0\end{matrix}$	$\begin{matrix}0\\1\end{matrix}$	$\begin{matrix}1\\0\end{matrix}$	$\begin{matrix}1\\1\end{matrix}$
q_0	$\left(q_0, \begin{matrix}0 & L\\ 0, & L\end{matrix}\right)$	$\left(q_0, \begin{matrix}1 & L\\ 1, & L\end{matrix}\right)$	$\left(q_0, \begin{matrix}1 & L\\ 0, & L\end{matrix}\right)$	$\left(q_1, \begin{matrix}0 & L\\ 1, & L\end{matrix}\right)$
q_1	$\left(q_0, \begin{matrix}1 & L\\ 0, & L\end{matrix}\right)$	$\left(q_0, \begin{matrix}0 & L\\ 1, & L\end{matrix}\right)$	$\left(q_1, \begin{matrix}0 & L\\ 0, & L\end{matrix}\right)$	$\left(q_1, \begin{matrix}0 & L\\ 1, & L\end{matrix}\right)$
q_R	$\left(q_R, \begin{matrix}0 & R\\ 0, & R\end{matrix}\right)$	$\left(q_R, \begin{matrix}0 & R\\ 1, & R\end{matrix}\right)$	$\left(q_R, \begin{matrix}1 & R\\ 0, & R\end{matrix}\right)$	$\left(q_R, \begin{matrix}1 & R\\ 1, & R\end{matrix}\right)$

rary result and the first factor copied to the second tape by the first Turing machine. An input configuration of tapes is shown in Figure 5.10 a). An output configuration of this machine is shown in Figure 5.10 b).

Let us make a few comments on states' purpose to clarify the design of the transition table shown in Table 5.6:

- q_0, q_1 – add numbers, heads simultaneously move left, corresponding digits of added numbers and the carry are added step by step,
- q_R – move heads right and place right of added numbers,
- r_0 – terminate computation and accept output data (when machines are joined, this state passes control to the second Turing machine).

The Turing machine linking the three ones discussed above is such as described below:

$$M = (Q_P \cup Q_R \cup Q_Q, \{0, 1\}, \{0, 1, B\}, \delta, p_1, B, \{r_A\})$$

where the transition function δ brings together transition functions δ_P, δ_R and δ_Q of branched Turing machines.

5.2 Nondeterministic Turing machines

Nondeterministic model is a very important modification of Turing machines. The equivalence of deterministic and nondeterministic Turing machines is the most important conclusion drawn from the discussion in this Chapter.

As mentioned before, for any configuration of deterministic Turing machines, at most, one transition could be done. Unlike nondeterministic Turing machines allow for a choice between several transitions for some (or all) configurations. This means that the transition function may yield a set of possible transitions. In this section, we provide a definition of nondeterministic Turing machine in basic model. Then we prove the equivalence between nondeterministic Turing machines in basic model and multi-tape deterministic Turing machines. A study on the equivalence of different models of nondeterministic Turing machines is similar to an analogous study on deterministic Turing machines. For that reason, we skip over such a discussion.

Definition 5.17. Nondeterministic Turing machine in basic model is a system

$$M = (Q, \Sigma, \Gamma, \delta, q_0, B, F, C)$$

with components as follows:
- δ is the transition function: $\delta : Q \times \Gamma \to \bigcup_{k=0}^{\infty}(Q \times \Gamma \times \{L, R\})^k$ where $(Q \times \Gamma \times \{L, R\})^0$ stands for an undefined value of transition function and $(Q\times\Gamma\times\{L, R\})^k$ denotes the set of values of transition function consisting of k 3-tuples being transition descriptions $(p, X, D) \in Q \times \Gamma \times \{L, R\}$.
- descriptions of other components are given in Definition 5.1.

Note that values of transition function are sets of several transitions (descriptions of) rather than one transition (like it was in the case of deterministic Turing machines), that is, we may have a set of k possible transitions as a value of the transition function:

$$\delta(q, X) = \{(p_1, Y_1, D_1), (p_2, Y_q, D_2), \ldots, (p_k, Y_k, D_k)\}$$

We will informally use a term that *transition function yields k values.*

Definition 5.18. Once the empty set or a one-element set $\{(p, Y, D)\}$ is a value of the transition function, then the transition is performed like for a deterministic machine, that is:
1. if the transition function yields the empty set, that is, it is undefined, then machine falls into infinite computation,
2. for a value $\{(p, Y, D)\}$ of the transition function, the control unit switches to state p, the head stores the symbol Y and shifts in direction D.

However, when the transition function yields a set of $k > 1$ values, then a transition could be explicated as follows:
3. a choice is made between possible k values,
4. for the chosen value, a transition is made like for deterministic Turing machine,
5. it is assumed that a transition chosen in the third point leads to such a computation (in the sense of deterministic Turing machines), which terminates in an accepting state if such a computation exists.

A question can be raised, how to hold the assumption of point 5, when a choice is made. Nondeterminism does not answer this question; it just assumes such choices.

Remark 5.4. The assumption made in point 5 of the above description is the fundamental assumption of nondeterminism. It creates an interpretation of nondeterminism, assuming that computation is a sequence of configurations achieved with *correct* choices between possible transitions.

Remark 5.5. Another interpretation of nondeterminism is reasonably practical. When the transition function yields a set of $k > 1$ values, the machine creates k copies of itself. Then, each copy makes a transition matching to one value and continues its own computation. The machine accepts if and only if such a copy is created during computation, which terminates its own computation in an accepting state.

In the spirit of the last interpretation, a nondeterministic Turing machine would be interpreted as a group of Turing machines. This group includes the original machine at the start of computation and might be enlarged during computation. Every machine of this group creates copies of itself, as many copies as the number of values yielded by the transition function. Then each copy makes a transition, as described above. Note that each copy is doing computation like a deterministic Turing machine.

Let us notice that the configuration (step description) of a nondeterministic Turing machine is exactly the same as for a deterministic one (of the same model). However, notions of transition relation and computation must be re-defined for nondeterministic Turing machines. Below, we discuss a (practical) interpretation of computation in the form of a tree representing all possible choices of values of a transition function.

Definition 5.19. A pair of configurations of a nondeterministic Turing machine is in the transition relation if and only if the second configuration can be derived from the first one by application of a transition yielded by the transition function, cf. Definition 5.3.

Definition 5.20. An interpretation of a computation of a Turing machine $M = (Q, \Gamma, \Sigma, \delta, q_0, B, F, C)$ is a tree such that:
- its nodes are labeled by configurations of the machine,
- its root is labeled by the initial configuration,
- for any node η_p, its each child η_c is related to it in the transition relation, that is, $\eta_p \succ \eta_c$.

Note that a computation tree is a k-tree, where k is the maximal number of values yielded by the transition function for given arguments. We say that degree of nondeterminism is k.

The computation of a nondeterministic Turing machine is a tree, which may have finite paths from the root to leaves and infinite paths beginning in the root. Interpreting a nondeterministic Turing machine as a group of its copies, we can associate finite paths with corresponding machine copies. Based on this interpretation, we can say that a nondeterministic Turing machine accepts its input if and only if there is a copy of the machine, which terminates its computation in an accepting state.

Definition 5.21. A nondeterministic Turing machine accepts its input if and only if the computation tree has a path from the root to a leaf labeled by a configuration, for which the control unit is in an accepting state.

Example 5.6. Design a Turing machine accepting the language $L = \{ww : w \in \{0,1\}^*\}$.

Solution. We design a nondeterministic Turing machine with halting accepting state and with guard. It works according to an algorithm briefly described as follows:

1. remember (store information in states) and mark the first symbol as belonging to the first half of the input word. This symbol begins the first half of the input word,
2. pass the head right, looking for the beginning of the second half of the input word. Since the second half of the word must begin with the symbol matching the remembered one, only such symbols are considered during the pass. When such a symbol is met, the machine makes a nondeterministic choice:
   a. either it is in the first half of the input word, or
   b. this symbol begins the second half of the input word,
3. continue the pass for point 2a,
4. mark the symbol as a symbol of the second half of the input word for point 2b,
5. shift left the head to the first unmarked symbol of the first half of the input word,
6. remember and mark the symbol as belonging to the first half of the input word,
7. pass the head right over the first half and over symbols marked as belonging to the second half of the input word,
8. mark the symbol as belonging to the second half of the input word if it matches the remembered one,
9. perform points 5–8 as long as there are unmarked symbols in both halves,
10. accept if and only if symbols of both parts are marked during the same pass left to right,
11. reject in any other case.

The following nondeterministic Turing machine implements the above algorithm:

$$M = (Q, \Sigma, \Gamma, \delta, q_0, B, \#, \{q_A\})$$

where:

- $Q = \{q_0, q_0^0, q_0^1, q_1, q_2^0, q_2^1, q_2, q_3^0, q_3^1, q_4, q_5, q_A, \}$ – the set of states,
- $\Gamma = \{0, 1, X, Y, \#, B\}$ – the tape alphabet,
- $\Sigma = \{0, 1\}$ – the input alphabet,
- δ – the transition function given in Table 5.7.

Note that the transition function is nondeterministic because some entries of its table hold more than one value. For this Turing machine, the maximal number of values yielded by the transition function is 2, that is, the degree of nondeterminism is 2.

Table 5.7: The transition function of the nondeterministic Turing machine designed in Example 5.6.

δ	0	1	X	Y	#	B
q_0	(q_0^0, X, R)	(q_0^1, X, R)				(q_5, B, L)
q_0^0	$(q_0^0, 0, R)$ (q_1, Y, L)	$(q_0^0, 1, R)$				
q_0^1	$(q_0^1, 0, R)$	$(q_0^1, 1, R)$ (q_1, Y, L)				
q_1	$(q_1, 0, L)$	$(q_1, 1, L)$	(q_2, X, R)	(q_1, Y, L)		
q_2	(q_2^0, X, R)	(q_2^1, X, R)		(q_4, Y, R)		
q_2^0	$(q_2^0, 0, R)$	$(q_2^0, 1, R)$		(q_3^0, Y, R)		
q_2^1	$(q_2^1, 0, R)$	$(q_2^1, 1, R)$		(q_3^1, Y, R)		
q_3^0	(q_1, Y, L)			(q_3^0, Y, R)		
q_3^1		(q_1, Y, L)		(q_3^1, Y, R)		
q_4				(q_4, Y, R)		(q_5, B, L)
q_5			(q_5, B, L)	(q_5, B, L)	$(q_A, \#, R)$	

Let us comment on the construction of the machine describing the purpose of states:

- q_0 – check if the input word is empty. If not, remember the input symbol in the state q_0^0 or q_0^1 and mark with X the first symbol of the word as belonging to the first half of the word,
- q_0^0 and q_0^1 – pass the head right over the first half of the word, nondeterministically find the first symbol of the second half of the word matching the remembered one, mark the found symbol with Y as belonging the second half of the word, switch to q_1 to pass the head left,
- q_1 and q_2 – pass the head to the first unmarked symbol of the first half of the word,
- q_2 – the next symbol of the word, as belonging to the first half of the word, is remembered in the state q_2^0 or q_2^1 and marked with X,
- q_2^0 and q_2^1 – the head is passed right over the first half of the word when the first symbol marking the second half of the word is found, states q_2^0 or q_2^1 are changed to q_3^0 or q_3^1 and pass of the head to the right is continued. Note that the middle of the word has already been marked and there is no need for using nondeterminism,
- q_3^0 and q_3^1 – the head is passed over the marked part of the second half of the word if the first unmarked symbol of the second half of the word matches the remembered one, the matching symbol is marked, the control unit switches to q_1 and the head shifts left,
- q_4 – when there are no unmarked symbols in the first half of the word, the head is passed right to check if there is no unmarked symbol in the second half of the word,
- q_5 – when both halves match, the head is passed left cleaning the tape (cleaning procedure), computation is terminated and input accepted,
- infinite computation begins in all configurations not considered above.

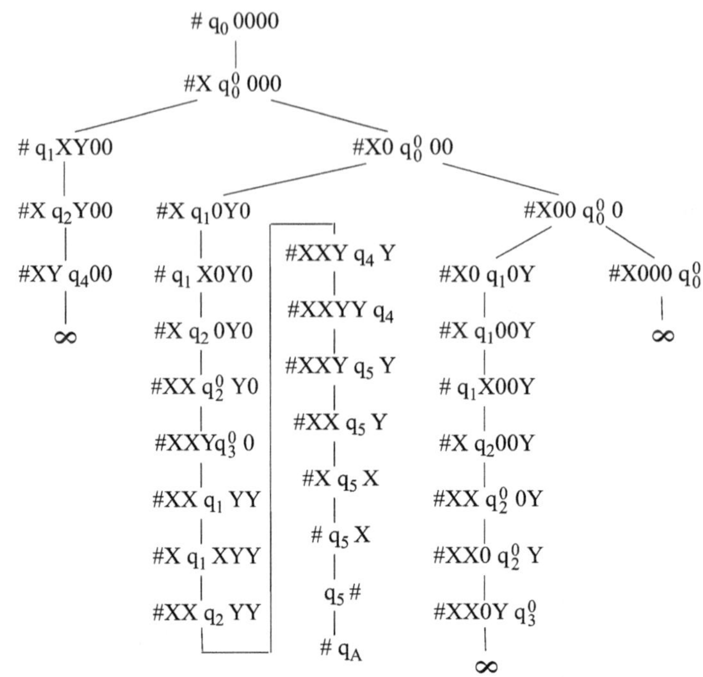

Figure 5.11: Computation of the nondeterministic Turing machine of Example 5.6 for the input word w = 0000.

An example of computation is shown in Figure 5.11. There are four paths in the computation tree. Three of them ended with the symbol of infinity which stands for infinite computation. The second path (from left) stands for computation terminated in the halting accepting state. Therefore, the input word is accepted by the machine.

For the sake of clarity, the transition table of the designed machine exposed in Example 5.7 has empty entries. The machine can be turned into a Turing machine with the stop property. It requires introducing the halting rejecting state and a few states for the cleaning procedure. In the transition table, all empty entries should be filled in with a switch to cleaning procedure (different than the one used for accepted word) and then switching to the halting rejecting state. On the other hand, configurations of the machine corresponding to empty entries of the transition table can never be reached.

Example 5.7. Rebuild the Turing machine designed in Example 5.6 to the one having the stop property.

Solution. The following nondeterministic Turing machine with halting states and with guard implements the above algorithm:

$$M = (Q, \Sigma, \Gamma, \delta, q_0, B, \#, \{q_A\}, \{q_R\})$$

Table 5.8: The transition function of the nondeterministic Turing machine with halting states, cf. Example 5.7. Empty entries of this table hold (q_6, B, R), which is not displayed for the sake of clarity.

δ	0	1	X	Y	#	B
q_0	(q_0^0, X, R)	(q_0^1, X, R)				(q_5, B, L)
q_0^0	$(q_0^0, 0, R)$ (q_1, Y, L)	$(q_0^0, 1, R)$				
q_0^1	$(q_0^1, 0, R)$	$(q_0^1, 1, R)$ (q_1, Y, L)				
q_1	$(q_1, 0, L)$	$(q_1, 1, L)$	(q_2, X, R)	(q_1, Y, L)		
q_2	(q_2^0, X, R)	(q_2^1, X, R)		(q_4, Y, R)		
q_2^0	$(q_2^0, 0, R)$	$(q_2^0, 1, R)$		(q_3^0, Y, R)		
q_2^1	$(q_2^1, 0, R)$	$(q_2^1, 1, R)$		(q_3^1, Y, R)		
q_3^0	(q_1, Y, L)			(q_3^0, Y, R)		
q_3^1		(q_1, Y, L)		(q_3^1, Y, R)		
q_4	$(q_6, 0, R)$	$(q_6, 1, R)$		(q_4, Y, R)		(q_5, B, L)
q_5			(q_5, B, L)	(q_5, B, L)	$(q_A, \#, R)$	
q_6	$(q_6, 0, R)$	$(q_6, 1, R)$				(q_7, B, L)
q_7	(q_7, B, L)	(q_7, B, L)	(q_7, B, L)	(q_7, B, L)	$(q_R, \#, R)$	(q_7, B, L)

where:

- $Q = \{q_0, q_0^0, q_0^1, q_1, q_2^0, q_2^1, q_2, q_3^0, q_3^1, q_4, q_5, q_6, q_7, q_A, q_R\}$ – the set of states,
- $\Gamma = \{0, 1, X, Y, \#, B\}$ – the tape alphabet,
- $\Sigma = \{0, 1\}$ – the input alphabet,
- δ – the transition function is given in Table 5.8.

Note that the new state q_R is the halting rejecting state and new states q_6 and q_7 are used in the cleaning procedure for the rejected input word.

Proposition 5.8. *The class of nondeterministic Turing machines is equivalent to the class of deterministic Turing machines.*

Proof. We will show that for a given Turing machine of one type, an equivalent machine of another type can be designed.

Note that a deterministic Turing machine in basic model is also a nondeterministic one (with the maximal number of values yielded by transition function not greater than 1, that is, with the degree of nondeterminism equal to one). For that reason and due to the equivalence of different types of Turing machines, any deterministic Turing machine is equivalent to some nondeterministic one.

Now, for a given nondeterministic Turing machine in basic model, we can design an equivalent deterministic multi-type Turing machine.

Let us present an idea of designing a deterministic Turing machine equivalent to a nondeterministic one. The idea is based on a deterministic simulation of a computation of a nondeterministic Turing machine. It can be briefly presented as follows:

- a nondeterministic Turing machine accepts its input if and only if the computation tree has a node, which terminates computation in an accepting state,
- a computation tree is a k-tree, where k is the maximal number of values yielded by the transition function, that is, degree of nondeterminism is k,
- a breadth-first tree searching algorithm, which starts searching from the root and, then visits nodes level by level, will eventually visit a node, which terminates computation in an accepting state,
- a simulation of a computation of a nondeterministic machine is rooted in the breadth-first searching the computation tree. This simulation investigates nodes visited by breadth-first search,
- for each visited node, the computation from the root to this node is reproduced. If this computation terminates in an accepting state, then the simulation is terminated with acceptation,
- otherwise, breadth-first searching is continued and a next node is investigated.

Note that the above method replicates transitions many times: closer a node to the root, more replications of the path from the root to this node is done. The time complexity of such a simulation is huge. Unfortunately, a more efficient deterministic method to simulate nondeterminism is not known.

The above idea can be realized by a 3-tape deterministic Turing machine. Let us assume the degree of nondeterminism is r. The simulation algorithm could be briefly described as follows:

1. an input of a nondeterministic machine is stored on the first tape,
2. the simulating deterministic machine systematically generates all possible sequences of length being successive natural numbers $0, 1, 2, \dots$. Elements of generated sequences are natural numbers from the interval $[1, 2, \dots, r]$. Each generated sequence is stored on the second tape,
3. for each generated sequence $(i_1, i_2, \dots, i_n)$ the following computation is done:
   - the input is copied from the first tape to the third one,
   - a path from the computation tree of a nondeterministic Turing machine is simulated. The path begins in the root of the computation tree and is defined by the sequence $(i_1, i_2, \dots, i_n)$:
     - length of the path is equal to the length of the sequence, that is, n transitions are simulated,
     - the successive l-th transition is defined by the i_l-th value of the transition function of the nondeterministic Turing machine. If the transition function yields less than k values for given arguments and the i_l-th value has no corresponding transition, then the simulation is terminated and the al-

gorithm goes to generate and investigate the next sequence of numbers, as described in point 2,

- if, for a given generated sequence, a node terminating computation in accepting state of the nondeterministic Turing machine is reached, then simulating machine halts its computation and accepts the input. Otherwise, the algorithm generates and investigates the next sequence of numbers, as described in point 2.

The above simulation algorithm describes a deterministic Turing machine, which is equivalent to the simulated nondeterministic Turing machine. □

5.3 Linear bounded automata

We can observe that the Turing machine built, for instance, in Example 5.6 exploits only this part of its tape during computation, which was used to store input data. We can consider and explore a class of Turing machines with this restriction. Indeed, such cases are already considered: Turing machines with stop property, which compute only on the part of their tape storing input word, have been studied. Such subclass of Turing machines is distinguished and it is called linear bounded automata. It occurs that linear bounded automata are equivalent to the class of context-sensitive grammars; that is, they accept the class of context-sensitive languages. The formal definition of linear bounded automata is as follows:

Definition 5.22. A linear bounded automaton in basic model is a Turing machine with the stop property:

$$M = (Q, \Sigma, \Gamma, \delta, q_0, B, \#, \&, F, C)$$

where:

- $\#, \&$ are the left and the right guards, $\#, \& \in \Gamma$, $\#, \& \notin \Sigma$,
- $\# q_0\, a_1\, a_2 \ldots a_n\, \&$ is the initial configuration, where $a_1\, a_2 \ldots a_n$ is input data,
- the transition function cannot yield a value that allows the head:
  - to replace the left guard symbol with any other symbol or to make a shift left when it reads the left guard symbol; that is, it may take only the following values $\delta(q, \#) = \{(p_1, \#, R), (p_2, \#, R), \ldots, (p_k, \#, R)\}$ for any $q, p_1, p_2, \ldots, p_k \in Q$,
  - to replace the right guard symbol with any other symbol or to make a shift right when it reads the right guard symbol, that is, it may take only the following values
    $\delta(q, \&) = \{(p_1, \&, L), (p_2, \&, L), \ldots, (p_k, \&, L)\}$ for any $q, p_1, p_2, \ldots, p_k \in Q$,
  - to store a guard symbol in any cell besides these holding them, that is
    if $X \in \Gamma - \{\#, \&\}$ and $q \in Q$,
    then $\delta(q, X) = \{(p_1, Y_1, D_1), (p_2, Y_2, D_2), \ldots, (p_k, Y_k, D_k)\}$,
    where $Y_1, Y_2, \ldots, Y_k \in \Gamma - \{\#, \&\}$, $p_1, p_2, \ldots, p_k \in Q$ and $D_1, D_2, \ldots, D_k \in \{L, R\}$,
- other components are as in Definition 5.1.

Notice that we can talk about two classes of linear bounded automata: deterministic and nondeterministic ones.

The discussion on varieties of Turing machines could hardly be adapted to the analysis of linear bounded automata. First of all, the notion of guards is directly adapted. We can talk about halting accepting and rejecting states and about a multi-track tape. Furthermore, these notions are easily adaptable to linear bounded automata. However, the nature of most models of Turing machines does not match the character of linear bounded automata. For instance, one-way and two-way infinite tape do not have counterparts in the class of linear bounded automata. Also, we do not discuss multi-tape linear bounded automata, though, formally, we can discuss a multi-tape restricted Turing machine such that its all tapes have a length equal to the length of input data limited by the left and the right guard.

There is a model, which raised the name of linear bounded automata. In this model length of the tape is bounded by a linear function of the length of the input data or, equivalently, is a multiply of the length of the input data. In these automata, the beginning and the end of the tape are marked by guards, as in basic model.

Proposition 5.9. *The following classes of deterministic linear bounded automata are equivalent:*
1. *in basic model,*
2. *with halting states,*
3. *with a multi-track tape,*
4. *with the length of the tape bounded by a linear function.*

The same classes of nondeterministic linear bounded automata are equivalent as well.

Proof. Proof of equivalence of these classes is quite similar to proofs of equivalence of corresponding classes of Turing machines.

Hint: Consider automaton with k-tracks tape of lengths equal to the length of an input word, which is equivalent to an automaton with one track tape k times longer than an input word. □

Deterministic vs. nondeterministic linear automata

It is trivial that for each deterministic linear bounded automaton, there exists an equivalent nondeterministic one. It is sufficient to turn (single) values of deterministic transition function to sets being values of nondeterministic transition function including every single value into the corresponding set. With regard to opposite equivalence, we do not know if there exists a deterministic linear bounded automaton for a given nondeterministic one. We know that it is possible to design a deterministic square bounded automaton (using part of tape of length proportional to square of input length), which is equivalent to a nondeterministic linear automaton.

Table 5.9: The transition function of the linear bounded automata designed in Example 5.8. Empty entries of the transition table keep transitions to the halting rejecting state. These entries are left empty due to the sake of clarity of the transition table.

δ	a	b	A	B	#	&	Λ
q_0	(q_A, A, R)	(q_0, b, R)	(q_0, A, R)	(q_0, B, R)		$(q_2, \&, L)$	
q_A	(q_A, a, R)	(q_A, b, R)	(q_A, A, R)	(q_A, B, R)		$(q_1, \&, L)$	
q_1	(q_1, a, L)	(q_B^1, b, L)	(q_1, A, L)	(q_1, B, L)			
q_B^1	(q_B^1, a, L)	(q_B^0, B, L)	(q_B^1, A, L)	(q_B^1, B, L)			
q_B^0	(q_B^0, a, L)	(q_B^1, b, L)	(q_B^0, A, L)	(q_B^0, B, L)	$(q_0, \#, R)$		
q_2		(q_3, Λ, L)	(q_2, Λ, L)	(q_2, Λ, L)			
q_3			(q_3, Λ, L)	(q_3, Λ, L)	$(p_A, \#, R)$		

Example 5.8. Design a linear bounded automaton accepting the language

$$L = \{w \in \{a, b\}^* : \#_a w = n, \#_b w = 2^n \text{ and } n \geq 0\}$$

Solution. Let us provide a linear bounded automaton with halting states:

$$M = (Q, \Sigma, \Gamma, \delta, q_0, \Lambda, \#, \&, \{p_A\}, \{p_R\})$$

where:

- $Q = \{q_0, q_A, q_1, q_B^0, q_B^1, q_2, q_3, p_A, p_R\}$ – the set of states,
- $\Gamma = \{a, b, A, B, \#, \&, \Lambda\}$ – the tape alphabet,
- $\Sigma = \{a, b\}$ – the input alphabet
- δ – the transition function given in Table 5.9,
- Λ – the blank symbol,
- p_A – the accepting state.

An algorithm applied in the automaton is as follows:

- the head is passed right and left along the tape,
  - during the walk right one letter a is marked,
  - during the walk left, half of the letters b are marked; that is, the number of letters b is divided by 2,
- passing the head right and left is continued until all letters a are marked, as a result n letters a and $2^n - 1$ letters b are marked,
- the final pass right checks that there are no letters a, the final pass left checks that there is only one letter b.

Let us comment on the construction of the automaton describing the purpose of states:

- q_0 – pass the head right from the beginning of the tape to its end, mark the first letter a met and indicate it switching to q_A; if no letter a is met, begin cleaning process leading to the acceptation of input data,
- q_A – continue passing the head right, memorize that the letter a was marked,
- q_1 – pass the head left from the end of the tape to its beginning, look for the first letter b, remember it switching to q_B^1,
- q_B^1 – continue passing the head left, look for the second letter b, mark it when met and remember it switching to q_B^0,
- q_B^0 – continue passing the head left, look for the next letter b, remember it switching to q_B^1,
- q_1, q_B^1, q_B^0 – count pairs of letters b, mark every second letter b (divide number of letters b by 2),
- q_2 and q_3 – pass the head left from the end of the tape to its beginning, mark the last letter b, terminate computation switching to the accepting state p_A.

Computations for $w_1 = bb \notin L$, $w_2 = b \in L$ and $w_3 = bbaabb \in L$ are shown below. Note that for the word w_1 the automaton rejects input data and that a cleaning procedure is not implemented due to the sake of clarity:

- $\# q_0\, b\, b\, \& \succ \# b\, q_0\, b\, \& \succ \# b\, b\, q_0\, \& \succ \# b\, q_2\, b\, \& \succ \# q_3\, b\, \Lambda\, \& \succ q_R\, \# \Lambda\, \Lambda\, \& \succ \# \Lambda\, q_R\, \Lambda\, \&$
- $\# q_0\, b\, \& \succ \# b\, q_0\, \& \succ \# q_2\, b\, \& \succ q_3\, \# \Lambda\, \& \succ \# p_A\, \Lambda\, \&$
- $\# q_0\, b\, b\, a\, a\, b\, b\, \& \succ \# b\, q_0\, b\, a\, a\, b\, b\, \& \succ \# b\, b\, q_0\, a\, a\, b\, b\, \& \succ \# b\, b\, A\, q_A\, a\, b\, b\, \& \succ$
  $\# b\, b\, A\, a\, q_A\, b\, b\, \& \succ \# b\, b\, A\, a\, b\, q_A\, b\, \& \succ \# b\, b\, A\, a\, b\, b\, q_a\, \& \succ \# b\, b\, A\, a\, b\, q_1\, b\, \& \succ$
  $\# b\, b\, A\, a\, q_B^1\, b\, b\, \& \succ \# b\, b\, A\, q_B^0\, a\, B\, b\, \& \succ \# b\, b\, q_B^0\, A\, a\, B\, b\, \& \succ \# b\, q_B^0\, b\, A\, a\, B\, b\, \& \succ$
  $\# q_B^1\, b\, b\, A\, a\, B\, b\, \& \succ q_B^0\, \# B\, b\, A\, a\, B\, b\, \& \succ \# q_0\, B\, b\, A\, a\, B\, b\, \& \succ \# B\, q_0\, b\, A\, a\, B\, b\, \& \succ$
  $\# B\, b\, q_0\, A\, a\, B\, b\, \& \succ \# B\, b\, A\, q_0\, a\, B\, b\, \& \succ \# B\, b\, A\, A\, q_A\, B\, b\, \& \succ^2 \# B\, b\, A\, A\, B\, b\, q_A\, \& \succ$
  $\# B\, b\, A\, A\, B\, q_1\, b\, \& \succ \# B\, b\, A\, A\, q_B^1\, B\, b\, \& \succ^3 \# B\, q_B^1\, b\, A\, A\, B\, b\, \& \succ$
  $\# q_B^0\, B\, B\, A\, A\, B\, b\, \& \succ q_B^0\, \# B\, B\, A\, A\, B\, b\, \& \succ \# q_0\, B\, B\, A\, A\, B\, b\, \& \succ^6$
  $\# B\, B\, A\, A\, B\, b\, q_0\, \& \succ \# B\, B\, A\, A\, B\, q_2\, b\, \& \succ \# B\, B\, A\, A\, q_3\, B\, \Lambda\, \& \succ^5$
  $q_3\, \# \Lambda\, \Lambda\, \Lambda\, \Lambda\, \Lambda\, \Lambda\, \& \succ \# p_A\, \Lambda\, \Lambda\, \Lambda\, \Lambda\, \Lambda\, \Lambda\, \&$

5.4 Problems

Problem 5.1. Design a Turing machine accepting the language: $L = \{w \in \{a, b, c\}^* : 0 \le \#_a w \le \#_b w \le \#_c w\}$

Solution. Let us provide a linear bounded automaton with accepting states:

$$M = (Q, \Sigma, \Gamma, \delta, q, B, \#, \&, \{p_A\}, \{p_R\})$$

where:

- $Q = \{q, q_a, q_b, q_c, q_{ab}, q_{ac}, q_{bc}, q_{abc}, p_L, p_T, p_A, p_R\}$ – the set of states,
- p_A and p_R – the accepting and rejecting states,

Table 5.10: The transition function of the linear bounded automata designed in Problem 5.1. Empty entries of the transition table keep transitions to the halting rejecting state: (q_R, B, D), where D defines shift to the left for the head not being over # or to the right for the head not being over &. These entries are left empty due to the sake of clarity of the transition table.

δ	a	b	c	B	#	&
q	(q_a, B, R)	(q_b, B, R)	(q_c, B, R)	$(q\ , B, R)$		$(q_T, \&, L)$
q_a	(q_a, a, R)	(q_{ab}, B, R)	(q_{ac}, B, R)	(q_a, B, R)		
q_b	(q_{ab}, B, R)	(q_b, b, R)	(q_{bc}, B, R)	(q_b, B, R)		
q_c	(q_{ac}, B, R)	(q_{bc}, B, R)	(q_c, c, R)	(q_c, B, R)		$(q_L, \&, L)$
q_{ab}	(q_{ab}, a, R)	(q_{ab}, b, R)	(q_{abc}, B, R)	(q_{abc}, B, R)		
q_{ac}	(q_{ac}, a, R)	(q_{abc}, B, R)	(q_{ac}, c, R)	(q_{ac}, B, R)		
q_{bc}	(q_{abc}, B, R)	(q_{bc}, b, R)	(q_{bc}, B, R)	(q_{bc}, B, R)		$(q_L, \&, L)$
q_{abc}	(q_{abc}, a, R)	(q_{abc}, b, R)	(q_{abc}, c, R)	(q_{abc}, B, R)		$(q_L, \&, L)$
q_L	(q_L, a, L)	(q_L, b, L)	(q_L, c, L)	(q_L, B, L)	$(q, \#, R)$	
q_T				(q_T, B, L)	$(p_A, \#, R)$	

- $\Gamma = \{a, b, c, \#, \&, B\}$ – the tape alphabet,
- $\Sigma = \{a, b, c\}$ – the input alphabet
- δ – the transition function given in Table 5.10,
- B – the blank symbol.

We use a linear bounded automaton with halting accepting state (which is a Turing machine) and apply the following algorithm to solve the problem:

- the head is passed from the beginning to the end of the tape, one letter a, b and c are marked, then the head is passed back to the beginning of the tape,
- the process of the previous point is continued until there are unmarked letters a, b and c or there are letters b and c or there is only letter(s) c,
- when no unmarked letters are found during the pass from the beginning to the end of the tape, then the head is passed back to the beginning of the tape and input is accepted,
- in all other cases the automaton rejects its input.

Let us comment the construction of the automaton describing purpose of states:

- $q, q_a, q_b, q_c, q_{ab}, q_{ac}, q_{bc}, q_{abc}$ – pass the head right from the beginning of the tape to its end, mark one letter a, b and c with the blank symbol and remember marked letters in a respective state,
- q_L – pass the head back to the beginning of the tape and then start passing the head right as in the previous point,
- continue the process of passing the head right and marking letters a, b and c until there are unmarked letters a, b and c or unmarked letters b and c or unmarked letter(s) c,
- q_T – pass the head back to the beginning of the tape and accept the input.

Note that the transition table may be simplified. The state q_{abc} and the row of the table corresponding to this state may be dropped. Instead, the value (q_{abc}, B, R) of transition function should be replaced by (q_L, B, L).

This optimization is not done in order to keep clearness of the process of passing the head from the beginning to the end of the tape.

And, finally, let us construct computations for words $w_1 = cbaabb \notin L$ and $w_3 = bbaccc \in L$, and are shown below. Note that for the word w_1, the automaton rejects the input data. In this case, a cleaning procedure is not implemented due to the sake of simplification of the solution.

- $\#q\ cbaab\& \succ \#Bq_c baab\& \succ \#BBq_{bc} aab\& \succ \#BBBq_{abc} ab\& \succ^2$
  $\succ \#BBBabq_{abc}\& \succ \#BBBaq_L b\& \succ^5 q_L\#BBBab\& \succ$
  $\succ \#q\ BBBab\& \succ^3 \#BBBq\ ab\& \succ^5 \#BBBBq_a b\& \succ$
  $\#BBBBBq_{ab}\& \succ \#BBBBp_R B\&$
- $\#q\ bbacc\& \succ \#Bq_b bacc\& \succ \#Bbq_b acc\& \succ \#BbBq_{ab} cc\& \succ$
  $\#BbBBq_{abc} c\& \succ \#XbXXcq_{abc}\& \succ \#BbBBq_L c\& \succ_5 q_L\#BbBBc\& \succ$
  $\#q\ BbBBc\& \succ \#Bq\ bBBc\& \succ \#BBq_b BBc\& \succ^2 \#BBBBq_b c\& \succ$
  $\#BBBBBq_{bc}\& \succ \#BBBBq_L B\& \succ^5 q_L\#BBBBB\& \succ \#q\ BBBBB\& \succ^5$
  $\#BBBBBq\ \& \succ \#BBBBq_T B\& \succ^5 q_T\#BBBBB\& \succ \#q_A BBBBB\&$

Problem 5.2. Design a Turing machine computing the function $f(x) = 2^x$ for a natural numbers x.

Problem 5.3. Design a Turing machine computing the function $f(x) = \log_2\lceil x\rceil$ for a positive natural numbers x.

Problem 5.4. Design a Turing machine converting numbers from the unary system to the binary positional system.

Problem 5.5. Design a Turing machine converting numbers from the binary positional system to the unary system.

Problem 5.6. Let $L_1 = L(M_1)$ and $L_2 = L(M_2)$ are languages accepted by deterministic Turing machines with the stop property M_1 and M_2. Build a Turing machine, which accepts:

- union of L_1 and L_2, that is, the language $L = L_1 \cup L_2$,
- intersection of L_1 and L_2, that is, the language $L = L_1 \cap L_2$.

Solution. The idea of designing a required Turing machine M is simple. M runs parallel computation of both machines M_1 and M_2 for a given input word. M accepts union $L = L_1 \cup L_2$ if and only if at least one machine of M_1 and M_2 accepts the input word. Intersection $L = L_1 \cap L_2$ is accepted by M if and only if both machines M_1 and M_2 accept the input word. In case that one machine terminates and another one still continues computation, the former one may perform empty *transitions*, that is, its state, head position and cell contents stay unchanged. For instance, when the union of languages is computed and one machine comes to the rejecting state, it is necessary to *wait* for termination of computation of another machine. Details of a design are given below.

Assume that M_1 and M_2 are given Turing machines with halting states and with two-way infinite tape

$$M_1 = (Q_1, \Sigma, \Gamma_1, \delta_1, q_0^1, B, \{q_A^1\}, \{q_R^1\})$$
$$M_2 = (Q_2, \Sigma, \Gamma_2, \delta_2, q_0^2, B, \{q_A^2\}, \{q_R^2\})$$

where $Q_1 \cap Q_2 = \varnothing$ and $\Gamma_1 \cap \Gamma_2 = \{B\}$.

A two-tape Turing machine is designed as follows

$$M = (Q_1 \times Q_2, \Sigma, \Gamma_1 \cup \Gamma_2, \delta, (q_0^1, q_0^2), B, (q_A^1, q_A^2), (q_R^1, q_R^2))$$

where an input word is copied to the second tape and heads of tapes are placed on the leftmost input symbol every tape. The transition function is defined on the basis of δ_1 and δ_2:
$\delta((q_1, q_2), (X_1, X_2)) = ((p_1, p_2), (Y_1, Y_2), (D_1, D_2))$, where
$\delta_1(q_1, X_1) = (p_1, Y_1, D_1), \delta_2(q_2, X_2) = (p_2, Y_2, D_2)$ for
$(q_1, q_2) \in (Q_1 - \{q_A^1, q_R^1\}) \times (Q_2 - \{q_A^2, q_R^2\})$,

- for union of : L_1 and L_2
  $\delta((q_R^1, q_2), (X_1, X_2)) = ((q_R^1, p_2), (X_1, Y_2), (S, D_2))$ and
  $\delta((q_1, q_R^2), (X_1, X_2)) = ((p_1, q_R^2), (Y_1, X_2), (D_1, S))$
  where $\delta_1(q_1, X_1) = (p_1, Y_1, D_1), \delta_2(q_2, X_2) = (p_2, Y_2, D_2)$,
  that is, if one machine out of M_1 and M_2 rejects, then M waits for termination of computation of another one,
  $\delta((q_A^1, q_2), (X_1, X_2)) = ((q_A^1, q_A^2), (X_1, X_2), (S, S))$ and
  $\delta((q_1, q_A^1), (X_1, X_2)) = ((q_A^1, q_A^2), (X_1, X_2), (S, S))$
  where $q_1 \in \Gamma_1$ and $q_2 \in \Gamma_2$,
  that is, if at least one machine out of M_1 and M_2 accepts then M accepts.
- for intersection of L_1 and L_2
  $\delta((q_A^1, q_2), (X_1, X_2)) = ((q_A^1, p_2), (X_1, Y_2), (S, D_2))$ and
  $\delta((q_1, q_A^2), (X_1, X_2)) = ((p_1, q_A^2), (Y_1, X_2), (D_1, S))$
  where $\delta_1(q_1, X_1) = (p_1, Y_1, D_1), \delta_2(q_2, X_2) = (p_2, Y_2, D_2)$,
  that is, if one machine out of M_1 and M_2 accepts, then M waits for termination of computation of another one,
  $\delta((q_R^1, q_2), (X_1, X_2), (D_1, D_2)) = ((q_R^1, q_R^2), (X_1, X_2), (S, S))$ and
  $\delta((q_1, q_R^1), (X_1, X_2), (D_1, D_2)) = ((q_R^1, q_R^2), (X_1, X_2), (S, S))$
  where $q_1 \in \Gamma_1$ and $q_2 \in \Gamma_2$,
  that is, M rejects if at least one machine out of M_1 and M_2 rejects.

Problem 5.7. Let $L = L(M)$ is a language accepted by a deterministic Turing machine with the stop property M. Build a Turing machine, which accepts complement of L, that is, the language $\overline{L} = \Sigma^* - L$, where Σ is an alphabet of L.

Solution. Assume that M is a given Turing machine with two-way infinite tape and with halting states:

$$M = (Q, \Sigma, \Gamma, \delta, q_0, B, \{q_A\}, \{q_R\})$$

The machine $\overline{M}$ with the accepting and rejecting states exchanged accepts the complement of L:

$$\overline{M} = (Q, \Sigma, \Gamma, \delta, q_0, B, \{q_R\}, \{q_A\})$$

Problem 5.8. Let $L_1 = L(M_1)$ and $L_2 = L(M_2)$ are languages accepted by deterministic Turing machines with the stop property M_1 and M_2. Build a Turing machine, which accepts concatenation of L_1 and L_2, that is, the language $L = L_1 \circ L_2$.

Solution. A solution is quite similar to the solution of Problem 5.6 for intersection. Instead of copying an input word to the second tape, its suffix is moved to the second tape. The suffix is nondeterministically chosen.

Problem 5.9. Let $L = L(M)$ is a language accepted by a Turing machine with the stop property M. Build a Turing machine, which accepts Kleene closure of L, that is, the language $L' = L^*$.

Solution. A nondeterministic two-tape Turing machine M' accepts the Kleene closure of $L(M)$. It realizes the following algorithm:
1. store an input word on the first tape,
2. if the first tape is empty, that is, it keeps the empty word ε, then M' accepts,
3. for a nonempty word at the input M' chooses nondeterministically a prefix of the current content of the first tape and moves the prefix to the second tape,
4. then M' runs the computation of M on the second tape,
5. if M rejects the copied prefix, then M' rejects its input,
6. switch to the second point of this algorithm.

For the sake of clarity, a cleaning procedure is not employed in this algorithm. A detailed description of M' is left to the reader.

Problem 5.10. Let $L_1 = L(M_1)$ and $L_2 = L(M_2)$ are languages accepted by Turing machines M_1 and M_2. Build a Turing machine, which accepts.
- union of L_1 and L_2, that is, the language $L = L_1 \cup L_2$,
- concatenation of L_1 and L_2, that is, the language $L = L_1 \circ L_2$,
- intersection of L_1 and L_2, that is, the language $L = L_1 \cap L_2$,
- Kleene closure of L_1, that is, the language $L = (L_1)^*$.

Hint: Adapt solutions of Problem 5.6, Problem 5.7, Problem 5.8 and Problem 5.9 to Turing machines without the stop property.

Problem 5.11. Adapt solutions of Problem 5.6, Problem 5.7, Problem 5.8 and Problem 5.9 for linear bounded automata instead of Turing machines with the stop property.

Hint: Consider solutions of this problems with regard to space used.

6 Pushdown automata

Pushdown automata vary from Turing machines in their definition and interpretation. However, despite differences, in this Chapter, we will prove that pushdown automata are limited Turing machines. Pushdown automata are finite structures with a stack, which is a potentially infinite element. A stack is a data structure, also called LIFO, that is, “last in, first out”, which allows for storing abstract elements of data and removing them. Usually two operations are used for operating a stack: *push* and *pop*. The *push* operation adds an element to the stack, hiding elements pushed earlier or initialized the stack if it is empty. The *pop* operation removes and returns the element most recently added to the stack or returns the empty value if the stack is empty (this is why stack is also called LIFO structure). Note that elements of the stack are not accessible except the one located on the top of the stack. To get access to a requested element formerly pushed on the stack is necessary to pop all elements pushed later than the requested one. In other words, elements are removed from the stack in the reverse order to the order they came. Thus, stack data accessibility is significantly limited comparing to tapes of Turing machines. As a result, pushdown automata are less powerful than Turing machines.

6.1 Nondeterministic pushdown automata

We discuss nondeterministic pushdown automata first. Deterministic ones should fulfill conditions, which are clearer on the basis of the general definition of nondeterministic automata.

Definition 6.1. A pushdown automaton is a system

$$M = (Q, \Sigma, \Gamma, \delta, q_0, \rhd, \lhd, F, R)$$

with components as follows:

Q a finite set of states,
Γ a finite set of stack symbols (stack alphabet),
$\rhd$ an initial stack symbol, $\rhd \in \Gamma$,
Σ a finite input alphabet,
$\lhd$ a special end-of-input symbol,
q_0 the initial state, $q_0 \in Q$,
F a set of accepting states, $F \subset Q$,
R a set of rejecting states, $R \subset Q, F \cap R = \varnothing$,
δ a transition function, $\delta : Q \times (\Sigma \cup \{\varepsilon\}) \times \Gamma \to \bigcup_{k=0}^{\infty} (Q \times \Gamma^*)^k$

A pushdown automaton could be interpreted as a physical mechanism shown in Figure 6.1. This mechanism consists of:

– a control unit, it is in a state $q \in Q$,

https://doi.org/10.1515/9783112223123-008

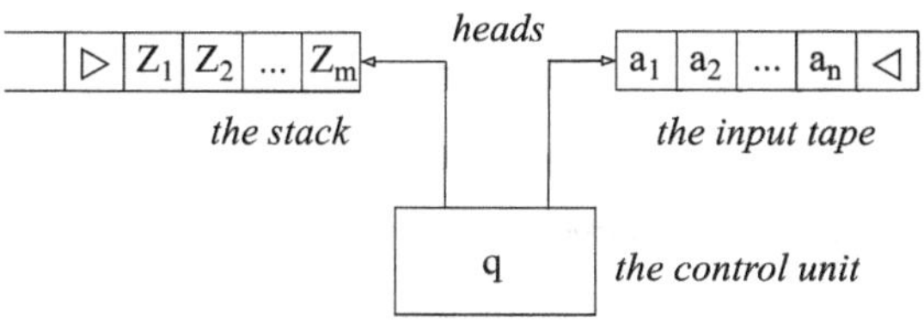

Figure 6.1: Pushdown automaton.

- an input tape holding input data $a_1 a_2 \dots a_n$,
- the special end-of-input symbol $\lhd$ is attached to input data, the symbol $\lhd$ neither belongs to the input alphabet nor to the stack alphabet, it marks the end of input easing practicing pushdown automata,
- an input head, which reads an input symbol and shifts right or does not take any action,
- if an input head reads end-of-input symbol $\lhd$, it does not shift,
- a stack, which is a one-way infinite tape with stack structure, which gives access only to its top cell, that is, the right most cell with Z_m symbol stored, see Figure 6.1.

A pushdown automaton is aimed at accepting a language, that is, answering if its input word belongs to the language or not. Computation of a given pushdown automaton is done according to the following intuitive procedure:
1. the initial configuration of a given pushdown automaton is described as follows:
   a. an input data, a word $w = a_1 a_2 \dots a_n$ over input alphabet Σ, is stored on the input tape, cf. Figure 6.1,
   b. the end-of-input symbol $\lhd$ is attached to the input data,
   c. the head of the input tape is placed over the first (leftmost) symbol of the input word,
   d. the head of the stack is (always) placed over the top cell of the stack
   e. the control unit is in the initial state q_0,
2. if the input head reads the end-of-input symbol $\lhd$ and the control unit is in an accepting state, then the computation is terminated and the automaton accepts the input,
3. if the control unit is in a rejecting state, then the computation is terminated and the automaton rejects the input. Notice that in this case, it is not required that the input head reads the end-of-input symbol $\lhd$. Of course, it would be possible to set such requirement, but better to be flexible for the sake of simplicity: the automaton responds to its state, that is, if a state of its control unit is an accepting one or a rejecting one,
4. if conditions of point 2 are not satisfied, then based on
   a. a state q of the control unit,
   b. a symbol X read by the head of the stack,
   c. either an input symbol a or without it,

the automaton makes the following actions:

d. values $\{(p_1, \alpha_1), (p_2, \alpha_2), \ldots, (p_k, \alpha_k)\}$ of the transition function $\delta(q, a, X)$ or $\delta(p, \varepsilon, X)$ are computed, where $a \in \Sigma$ is the input symbol, $X \in \Gamma$ is the top symbol of the stack, $q, p_1, p_2, \ldots, p_k \in Q$ are states, $\alpha_1, \alpha_2, \ldots, \alpha_k \in \Gamma^*$ are strings of symbols of the stack alphabet,
e. if values of the transition function are computed based on an input symbol $\delta(q, a, X)$, then the input head is shifted right, otherwise the input head does not change its position,
f. a value (p_i, α_i) of the transition function is chosen nondeterministically,
g. the top symbol X of the stack is removed and then symbols of the string α_i are pushed on the stack in reverse order, that is, the last symbol first, the first one last. The first symbol of α_i will be on the top of the stack after this operation,
h. the control unit switches to the state p_i,

5. computation goes to the point 2.

We assume that pushdown automata always terminate computation. The value $(Q \times \Gamma^*)^0$ of a transition function, which is the empty word ε, is interpreted as termination of computation in a rejecting state.

Note that pushdown automata cannot store output information, so then, unlike Turing machines, they can only accept languages and cannot compute functions or solve problems.

Now we define configuration (a step description), a transition relation and a computation of pushdown automata.

Definition 6.2. A configuration (a step description) of a pushdown automaton $M = (Q, \Sigma, \Gamma, \delta, q_0, \triangleright, \triangleleft, F, R)$ is the following sequence of symbols:

$$\gamma \, q \, w$$

where:
- $q \in Q$ is the current state of the control unit of a pushdown automaton,
- γ is the stack contents, the last symbol of γ is the top symbol of the stack,
- ω is the current input, the first symbol of w is the symbol under the input head, $\triangleleft$ is the last symbol of the input.

Note that both sequences γ and ω of symbols are words over the stack alphabet Γ and the input alphabet Σ and that any of these sequences may include only the initial stack symbol $\triangleright$ and the end-of-input symbol $\triangleleft$. However, none of these two sequences can be infinite, like in a case of a step description of a Turing machine.

For instance, the initial configuration is of the form $\triangleright \; q_0 \; w \; \triangleleft$, where q_0 is the initial state, w is the input data. In this case, the stack is empty, so the initial stack symbol is near the initial state symbol. On the other hand, a step description $\gamma \;\; q \; \triangleleft$ informs that input symbols have already been read.

Let us analyze transitions done by pushdown automata. We assume that the following sequence of symbols describes a step description:

$$\triangleright X_1\ X_2\ \dots\ X_{m-1}\ X_m\ \ q\ \ a_i\ a_{i+1}\ \dots\ a_n\ \triangleleft$$

Recall that:

- q is the current state of the control unit,
- $\triangleright X_1\ X_2\ \dots\ X_{m-1}\ X_m$ is the sequence of symbols on a stack, X_m is the top symbol of a stack,
- $a_i\ a_{i+1}\ \dots\ a_n \triangleleft$ is the sequence of input symbols, a_i is the front input symbol
- the head of a stack reads X_m,
- the input head either reads a_i, or does not read anything.

The transition function determines the next step description (configuration):

- if the value of the transition function is $\delta(q, a_i, X_m) = \{(p_1, \varepsilon), (p_2, Y_2^1\ Y_2^2\ Y_2^3)\}$, for instance, and the second value $\{(p_2, Y_2^1\ Y_2^2\ Y_2^3)\}$ is chosen nondeterministically, then the following step description is yielded,

$$\triangleright X_1\ X_2\ \dots\ X_{m-1}\ Y_2^3\ Y_2^2\ Y_2^1\ p_2\ a_{i+1}\ \dots\ a_n \triangleleft$$

- if the value of the transition function is $\delta(q, \varepsilon, X_m) = \{(p_1, \varepsilon), (p_2, Y_2^1\ Y_2^2\ Y_2^3)\}$, for instance, and the first value (p_1, ε) is chosen nondeterministically, then the following step description is yielded:

$$\triangleright X_1\ X_2\ \dots\ X_{m-1}\ p_1\ a_i\ a_{i+1}\ \dots\ a_n \triangleleft$$

We will use the symbol $\succ$ to denote a transition of a pushdown automaton. The transition symbol $\succ$ may be supplement with a pushdown automaton name $\succ_M$ to emphasize that a transition concerns a given pushdown automaton. It also can be supplemented with a superscript $\succ^k$ to notify that k transitions are done.

The above two transitions done by a pushdown automaton will be denoted as follows:

$$\triangleright X_1\ \dots\ X_{m-1}\ X_m\ q\ a_i\ a_{i+1}\ \dots\ a_n \triangleleft \succ \triangleright X_1\ \dots\ X_{m-1}\ Y_2^3\ Y_2^2\ Y_2^1\ p_2\ a_{i+1}\ \dots\ a_n \triangleleft$$
$$\triangleright X_1\ \dots\ X_{m-1}\ X_m\ q\ a_i\ a_{i+1}\ \dots\ a_n \triangleleft \succ \triangleright X_1\ \dots\ X_{m-1}\ p_2\ a_i\ a_{i+1}\ \dots\ a_n \triangleleft$$

Definition 6.3. Transitions of a pushdown automaton create a binary relation in the space of all possible configurations of the automaton, that is, any two configurations are related if and only if the second is derived from the first one by application of a transition function. This relation is called the transition relation of a given pushdown automaton. The transitive closure of the transition relation is denoted by $\succ^*$.

Now we define what is a computation of a pushdown automaton in the spirit of the corresponding description for nondeterministic Turing machines, cf. Remark 5.5. It is a practical interpretation of the computation.

Definition 6.4. An interpretation of a computation of a pushdown automaton is a tree such that:

- its nodes are labeled by configurations of an automaton,
- its root is labeled by the initial configuration,
- for any node η_p, its each child η_c is related to it in the transition relation, that is, $\eta_p \succ \eta_c$.

Nevertheless, a fundamental description of a computation of a pushdown automata is strictly consistent with the corresponding description for Turing machines outlined in Definition 5.18 and Remark 5.4.

Note that a computation tree is a $(k+l)$-tree, where k and l are the maximal number of values yielded by the transition function for given arguments. Note that a transition for a given state and a given stack symbol can be chosen from k-transitions respective to an input symbol and from l-transitions respective to so-called ε-transition, that is, when an input symbol is neither checked nor read.

Now we give a formal definition of acceptation of an input by a pushdown automaton.

Definition 6.5. A pushdown automaton accepts its input if and only if the pair of the initial configuration and a final configuration belongs to the transitive closure of a transition relation, that is, $\eta_1 \succ^* \eta_T$, where η_1 is an initial configuration and η_T is a final configuration.

Remark 6.1. A pushdown automaton accepts its input if and only if the computation tree has a path from the root to a leaf labeled by an accepting configuration.

Based on the above discussion, we now give formal definitions of some concepts.

Definition 6.6. The language accepted by a pushdown automaton is the set of words $w \in \Sigma^*$ accepted by that automaton.

Example 6.1. Design a pushdown automaton accepting the language of palindromes over the alphabet $\Sigma = \{a, b\}$ that is:

$$L = \{w \in \{a,b\}^* : w = w^R, \text{ where } w^R \text{ is reversed } w\},$$

Solution. First, we briefly comment on an algorithm for a pushdown automaton accepting this language:

- symbols corresponding to letters of the first part of an input word will be pushed on the stack,
- the middle of an input word is found nondeterministically; if an input word is of odd length, the middle letter is read without the stack change,
- letters of the second part of an input word are compared to stack symbols; if they match, the top symbol of the stack is popped out and the next pair of an input letter and a stack symbol are compared,

- an automaton accepts when both the input and the stack are empty,
- an automaton rejects in all other cases.

A pushdown automaton accepting this language:

$$M = (Q, \Sigma, \Gamma, \delta, p_1, \rhd, \lhd, F, R)$$

where:
- $Q = \{p_1, p_2, ACC, REJ\}$ – a set of states,
- $\Sigma = \{a, b\}$ – an input alphabet,
- $\Gamma = \{A, B, \rhd\}$ – a stack alphabet,
- p_1 – the initial state,
- $\rhd$ – the initial stack symbol,
- $\lhd$ – the end-of-input symbol,
- $F = \{ACC\}$ – a set of accepting states,
- $F = \{REJ\}$ – a set of rejecting states,
- $\delta : Q \times (\Sigma \cup \{\varepsilon\}) \times \Gamma \to (Q \times \Gamma^*)^0 \cup (Q \times \Gamma^*)^1 \cup (Q \times \Gamma^*)^2$ – a transitions function given in Table 6.1. Note that transition function depends on three arguments, so it cannot be given in one two-dimensional array. Thus, there are tables for every state.

Table 6.1: The transition function of the pushdown automaton designed in Example 6.1. Note, in order to make the table easy readable $\delta(p_i, X, Y) = (REJ, Y)$ is denoted by REJ. Similarly, $\delta(p_i, X, Y) = (ACC, Y)$ is denoted by ACC.

$\delta(p_1)$	a	b	ε	$\lhd$
A	(p_1, AA) (p_2, A)	(p_1, BA) (p_2, A)	(p_2, A)	REJ
B	(p_1, AB) (p_2, B)	(p_1, BB) (p_2, B)	(p_2, B)	REJ
$\rhd$	$(p_1, A\rhd)$ $(p_2, \rhd)$	$(p_1, B\rhd)$ $(p_2, \rhd)$	$(p_2, \rhd)$	ACC

$\delta(p_2)$	a	b	ε	$\lhd$
A	(p_2, ε)	REJ	REJ	REJ
B	REJ	(p_2, ε)	REJ	REJ
$\rhd$	REJ	REJ	REJ	ACC

Let us comment on the design of the automaton describing the purpose of the states:
- p_1 – accept the empty word at the input. If an input word is nonempty:
  - push on the stack symbols A or B corresponding to input letters a and b,
  - switch to p_2 when the middle of an input word is encountered, read the middle input letter when an input word is of odd length, otherwise switch without reading an input letter (make ε-transition),

- p_2 – read remaining input letters and pop symbols from the stack if they match, otherwise reject an input word,
- p_2 – accept an input word when the input and the stack are empty.

Examples of computation of the above pushdown automaton are given in Figure 6.2 and Figure 6.3.

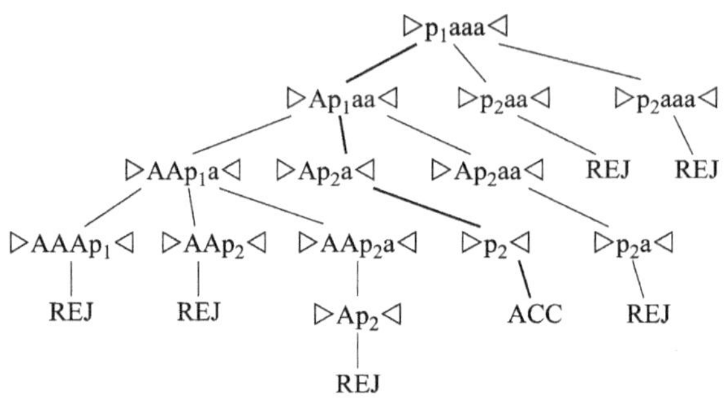

Figure 6.2: The computation of the pushdown automaton designed in Example 6.1 for the input word $w = aaa$.

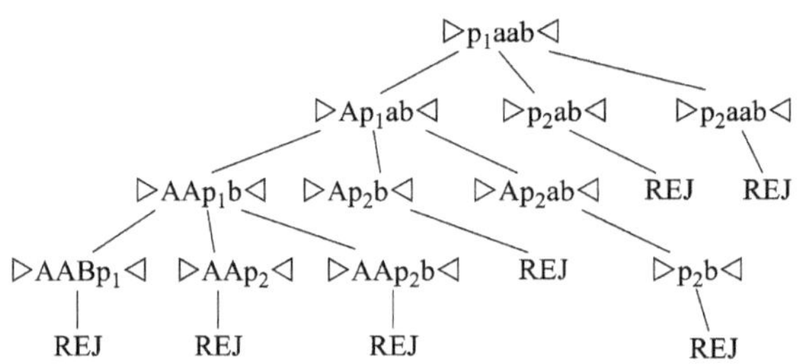

Figure 6.3: The computation of the pushdown automaton designed in Example 6.1 for the input word $w = aab$.

In Figure 6.2, a computation for the input word $w = aaa$ is given. The input word is accepted because there is a path from the root to an accepting leaf.

A computation for the input word $w = aab$, given in Figure 6.3, has no accepting leaf. As a result, the input word is rejected.

6.2 Deterministic pushdown automata

A pushdown automaton is deterministic if there is no more than one possible transition in any configuration. This condition needs that for any configuration of an automaton, a transition function yields at most one possible transition. However, for pushdown automata, there is another nondeterministic factor: for a given state and a given stack symbol, there might be a choice between transitions for given input symbols and without involving an input symbol, that is, so called ε-transition. As a result, the following definition formulates conditions for a pushdown automaton to be a deterministic one.

Definition 6.7. A pushdown automaton $M = (Q, \Sigma, \Gamma, \delta, q_0, \rhd, \lhd, F, R)$ is a deterministic one if and only if:

- its transition function for any arguments, that is, for any triple (q, X, Y), $q \in Q$, $X \in (\Sigma \cup \{\varepsilon, \lhd\})$, $Y \in \Gamma$, yields at most one transition,
- for given $q \in Q$ and $Y \in \Gamma$ the transition function rejects for (q, ε, Y) or rejects for all (q, a, Y), $a \in (\Sigma \cup \{\varepsilon\})$.

Proposition 6.1. *Deterministic pushdown automata are not equivalent to nondeterministic ones.*

Proof. It is intuitively clear that the language $L = \{w \in \{a, b\}^* : w = w^R\}$ discussed in Example 6.1 cannot be accepted by a deterministic pushdown automaton. On the other hand, deterministic pushdown automata are special cases of nondeterministic ones. From a perspective of accepted languages, the class of languages accepted by deterministic automata is included but not equal to the class of languages accepted by nondeterministic pushdown automata. A formal proof of strict inclusion can be found in [4]. □

Example 6.2. Design a pushdown automaton accepting the following language over the alphabet $\Sigma = \{a, b\}$:

$$L = \{w \in \{a, b\}^* : |\#_a w - \#_b w| \le 2\},$$

Solution. First, we briefly comment on an algorithm for a pushdown automaton accepting this language:

- numbers of letters a and b are compared in an input word; superfluous letters are pushed on the stack,
- number of symbols on the stack is checked when the input is empty; an input word is accepted if this number does not exceed 2.

A pushdown automaton accepting this language:

$$M = (Q, \Sigma, \Gamma, \delta, p_0, \rhd, \lhd, F, R)$$

where:

- $Q = \{p_0, p_1, p_2, \ ACC, \ REJ\}$ – a set of states,
- $\Sigma = \{a, b\}$ – an input alphabet,
- $\Gamma = \{A, B, \rhd\}$ – a stack alphabet,
- $\delta : Q \times \Sigma \times \Gamma \to Q \times \Gamma^*$ – a transitions function given in Table 6.2. Note that transition function neither yields multiply transitions, nor has ε-transitions. Therefore, the automaton is a deterministic one,
- other components of the automaton are like in Example 6.1.

Table 6.2: The transition function of the pushdown automaton designed in Example 6.2.

$\delta(p_0)$	a	b	$\lhd$
A	(p_0, AA)	(p_0, ε)	(p_1, ε)
B	(p_0, ε)	(p_0, BB)	(p_1, ε)
$\rhd$	$(p_0, A\rhd)$	$(p_0, B\rhd)$	ACC

$\delta(p_1)$	a	b	$\lhd$
A	REJ	REJ	(p_2, ε)
B	REJ	REJ	(p_2, ε)
$\rhd$	REJ	REJ	ACC

$\delta(p_2)$	a	b	$\lhd$
A	REJ	REJ	REJ
B	REJ	REJ	REJ
$\rhd$	REJ	REJ	ACC

The design of the automaton is commented in the form of the purpose of states descriptions:

- p_0 – numbers of letters a and b are compared in an input word; superfluous letters are pushed on the stack:
  - having a (or b) on the input, push the matching symbol A (or B) on the stack if it has the same symbol A (or B) on the top or it is empty (has $\rhd$ on the top),
  - pop the top symbol from the stack if it does not match an input symbol,
- p_0 – if the input is empty then:
  - accept if the stack is empty,
  - pop the top symbol from the stack and switch to p_1,
- p_1 – now the input is empty:
  - accept if the stack is empty,
  - pop the top symbol from the stack and switch to p_2,
- p_2 – accept if the stack is empty:
- reject in all other cases.

Note that states p_0, p_2 and p_1 check if the number of remaining symbols on the stack does not exceed 2.

Examples of computation of the automaton built in Example 6.2 for words $aaaa \notin L$ and $aaba \in L$ are given below:

$$\begin{aligned}
&\rhd p_0\, a\,a\,a\,a \lhd\ \succ\ \rhd A\, p_0\, a\,a\,a \lhd\ \succ\ \rhd A\,A\, p_0\, a\,a \lhd\ \succ\ \rhd A\,A\,A\, p_0\, a \lhd \\
&\qquad \succ\ \rhd A\,A\,A\,A\, p_0 \lhd\ \succ\ \rhd A\,A\,A\, p_1 \lhd\ \succ\ \rhd A\,A\, p_2 \lhd\ \succ\ REJ \\
&\rhd p_0\, a\,a\,b\,a \lhd\ \succ\ \rhd A p_0\, a\,b\,a \lhd\ \succ\ \rhd A\,A\, p_0\, b\,a \lhd\ \succ\ \rhd A\, p_0\, a \lhd \\
&\qquad \succ\ \rhd A\,A\, p_0 \lhd\ \succ\ \rhd A\, p_1 \lhd\ \succ\ \rhd p_2 \lhd\ \succ\ ACC
\end{aligned}$$

6.3 Accepting states versus empty stack

Examples presented in previous sections show that acceptance came with the initial symbol of the stack $\rhd$ and the end-of-input symbol $\lhd$. We may ask a question if this is a coincidence or rather a rule. That is, if we can change the acceptation by accepting state to the acceptation by the empty stack. Notice that the empty input is the default because we already assumed that acceptance must be accompanied by empty input. We can answer this question positively. We may move up a class of pushdown automata, which accept when the stack is empty (and, of course, the input is empty), and prove that automata accepting by empty stack are equivalent to automata accepting with an accepting state. An idea of the proof is quite clear.

On the one hand, having a pushdown automaton accepting with a state, we can empty its stack when an accepting state is reached and then accept its input. On the other hand, given a pushdown automaton accepting with an empty stack, it should make an additional transition to an extra accepting state when its stack is empty. Details are given below.

Definition 6.8. A pushdown automaton accepting by empty stack is a system

$$M = (Q, \Sigma, \Gamma, \delta, q_0, \rhd, \lhd, \varnothing, R)$$

where the set of accepting states is empty and other components are as shown in Definition 6.1.

Note that acceptance by empty stack does not depend on a state of a terminated automaton description. Because of this, the set of accepting states of such an automaton is empty.

Proposition 6.2. *Acceptance by states is equivalent to acceptance by empty stack.*

Proof. We prove that for any pushdown automaton accepting by accepting states, there exists an equivalent automaton accepting by empty stack and oppositely.

Assume that there is a pushdown automaton accepting by accepting states

$$M = (Q, \Sigma, \Gamma, \delta, q_0, \rhd, \lhd, F, R)$$

The following pushdown automaton accepting with the empty stack is equivalent to the above one

$$M' = (Q', \Sigma, \Gamma', \delta', q_0', \rhd', \lhd, \varnothing, R)$$

where:

1. $Q' = Q \cup \{q_0', q'\}$ and $Q \cap \{q_0', q'\} = \varnothing$,
2. $\Gamma' = \Gamma \cup \{\rhd'\}$ and $\Gamma \cap \{\rhd'\} = \varnothing$,
3. the transition function δ' is designed as follows:

a. $\delta'(q_0', \varepsilon, \rhd') = \{(q_0, \rhd\rhd')\}$,
b. $\delta'(q, a, X) = \delta(q, a, X)$ for $q \in Q$, $a \in \Sigma$, $X \in \Gamma$,
c. $\delta'(q, \varepsilon, X) = \delta(q, \varepsilon, X)$ for $q \in (Q - F)$, $X \in \Gamma$,
d. $\delta'(q, \varepsilon, X) = \{(q', \varepsilon)\}$ for $q \in F$, $X \in \Gamma - \{\rhd\}$ start emptying the stack,
e. $\delta'(q', \varepsilon, X) = \{(q', \varepsilon)\}$ for $X \in \Gamma - \{\rhd\}$ continue emptying the stack,
f. $\delta'(q', \lhd, \rhd) = \{(q', \varepsilon)\}$, removes the initial stack symbol and keeps the end-of-input symbol, accept with empty stack and empty input,
g. M' rejects for all other configurations.

The automaton M' moves to the initial configuration of M in its first transition, compare 3a. Then, it follows computation of M, see 3b and 3c. When M reaches an accepting configuration (recall, that it terminates computation for an accepting state), then M' pops a top symbol from its stack and switches to extra state q' in order to begin emptying the stack, compare 3d. Then, M' continues emptying the stack in 3d and finally it removes extra stack initial symbol $\rhd$ leaving empty its input and its stack, compare 3f. An extra stack initial symbol was needed in order to prevent confusion when the stack of the automaton M becomes empty during the computation of M.

It is worth noticing that the automaton M' is a nondeterministic one. Also, let us recall that accepting by switching to accepting state requires empty input, compare point 3 of computation description on page 131.

Now, we design a pushdown automaton accepting by states equivalent to a given one accepting by the empty stack. Assume that there is a pushdown automaton accepting with the empty stack

$$M = (Q, \Sigma, \Gamma, \delta, q_0, \rhd, \lhd, \varnothing, R)$$

The following pushdown automaton accepting with states is equivalent to the above one:

$$M' = (Q', \Sigma, \Gamma', \delta', q_0', \rhd', \lhd, F, R)$$

where:
- $Q' = Q \cup \{q_0', q_A\}$ and $Q \cap \{q_0', q_A\} = \varnothing$,
- $\Gamma' = \Gamma \cup \{\rhd'\}$ and $\Gamma \cap \{\rhd'\} = \varnothing$,
- $F = \{q_A\}$,
- the transition function δ' is design as follows:
  - $\delta'(q_0', \varepsilon, \rhd') = \{(q_0, \rhd\rhd')\}$,
  - $\delta'(q, a, X) = \delta(q, a, X)$ for $q \in Q$, $a \in \Sigma$, $X \in \Gamma$,
  - $\delta'(q, \varepsilon, X) = \delta(q, \varepsilon, X)$ for $q \in Q$, $X \in (\Gamma - \{\rhd\})$,
  - $\delta'(q, \varepsilon, \rhd) = \{(q_A, \varepsilon)\}$ for $q \in F$,
  - M' rejects for all other configurations.

The automaton M' goes to the initial configuration of M in its first transition; see the first point. Then, it follows a computation of M. When M empties its stack, then M' pops a top symbol from its stack (it is the initial symbol of M) and switches (nondeterministically) to the accepting state q_A and accepts if its input is empty. Otherwise, when input is not empty, M' continues (nondeterministically) simulation of M. □

6.4 Pushdown automata as Turing machines

Pushdown automata are restricted Turing machines. Moreover, since pushdown automata always terminate their computation, they are restricted Turing machines with the stop property.

Proposition 6.3. *There is a Turing machine with the stop property equivalent to a given pushdown automaton.*

Proof. Let us design a Turing machine equivalent to a given pushdown automaton accepting with states:

$$M = (Q, \Sigma, \Gamma, \delta, q_0, \triangleright, \triangleleft, F, R)$$

We design a 2-tape Turing machine with terminating states, which is equivalent to the automaton M:

$$M_T = (Q_T, \Sigma, \Gamma_T^1, \Gamma_T^2, \delta_T, B, \{q_A\}, \{q_R\})$$

where:

1. $Q_T = (Q - (F \cup R)) \cup \{q_A, q_R\} \cup Q_S$, $Q \cap (\{q_A, q_R\} \cup Q_S) = \emptyset$ – the set of states of the Turing machine M_T includes:
   a. states of the automaton M except accepting and rejecting states (except states, which terminate computation of the automaton),
   b. a new halting accepting state and a new halting rejecting state of the Turing machine M_T, and
   c. additional state Q_S, which simulate transitions of the automaton M,
2. $\Gamma_T^1 = \Sigma \cup \{\triangleleft, B\}$ – an alphabet of the first tape,
3. $\Gamma_T^2 = \Gamma \cup \{\triangleright, B\}$ – an alphabet of the second tape,
4. B – the blank symbol of both tapes.

A configuration of a pushdown automaton is characterized by the following configuration of the Turing machine, cf. Figure 6.4:

- an input is stored on the first tape, the head of the first tape is placed over the first input symbol,
- a stack is stored on the second tape; the initial stack symbol is the leftmost nonblank symbol of the tape, a top stack symbol is a rightmost nonblank tape symbol, the head of the tape is placed on a rightmost nonblank symbol (a top symbol of the stack),

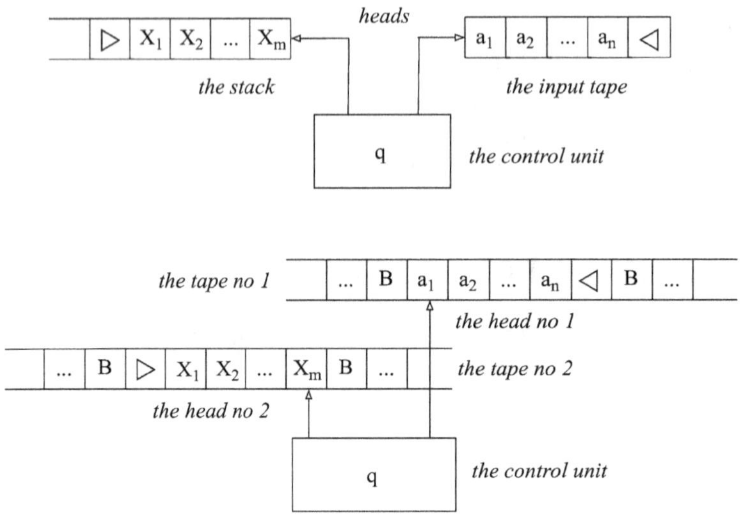

Figure 6.4: A pushdown automaton and its characterization by a Turing machine.

- the control unit of the Turing machine is in the same state as the control unit of the pushdown automaton.

All values of the transition function δ of the automaton are uniquely enumerated. Namely, if the transition function δ yields k transitions for a given triple of arguments: $\delta(q,a,X) = \{(p_1,\alpha_1),(p_2,\alpha_2),\dots,(p_k,\alpha_k)\}$ and this value of the transition has index t, then each pair (p_i,α_i) gets its own double index t and i.

For a given transition of the automaton $(p,\alpha) \in \delta(q,a,X)$ with a given number t and $p \in Q-(F \cup R)$, $a \in \Sigma \cup \{\triangleleft\}$ and $X \in \Gamma$, the Turing machine makes several transitions in order to update content of the second tape. All such transitions are uniquely identified by states corresponding to the given move of the pushdown automaton. The simulation of one transition of the automaton is formally described as follows:

1. if α is the empty sequence, then $(p, \genfrac{}{}{0pt}{}{B}{B}, \genfrac{}{}{0pt}{}{R}{L}) \in \delta'(q, \genfrac{}{}{0pt}{}{a}{X})$,
2. if $\alpha = X_1X_2\dots X_r$, then the set of states Q_T is expanded by adding states $p_1^t, p_2^t, \dots, p_r^t$ and the following transitions are included:
   a. $(p_r^t, \genfrac{}{}{0pt}{}{a}{X_r}, \genfrac{}{}{0pt}{}{S}{R}) \in \delta'(q, \genfrac{}{}{0pt}{}{a}{X})$,
   b. $\delta'(p_r^t, \genfrac{}{}{0pt}{}{a}{B}) = \{(p_{r-1}^t, \genfrac{}{}{0pt}{}{a}{X_{r-1}}, \genfrac{}{}{0pt}{}{S}{R})\}$,
   c. $\delta'(p_{r-1}^t, \genfrac{}{}{0pt}{}{a}{B}) = \{(p_{r-2}^t, \genfrac{}{}{0pt}{}{a}{X_{r-2}}, \genfrac{}{}{0pt}{}{S}{R})\}$,
   d. $\dots$
   e. $\delta'(p_2^t, \genfrac{}{}{0pt}{}{a}{B}) = \{(p_1^t, \genfrac{}{}{0pt}{}{a}{X_1}, \genfrac{}{}{0pt}{}{S}{R})\}$,
   f. $\delta'(p_1^t, \genfrac{}{}{0pt}{}{a}{B}) = \{(p, \genfrac{}{}{0pt}{}{B}{B}, \genfrac{}{}{0pt}{}{R}{L})\}$

A simulation of a transition $(p,\alpha) \in \delta(q,\varepsilon,X)$ is similar to simulation of a transition $(p,\alpha) \in \delta(q,a,X)$ with the two following changes:

- point 1 changes to: if α is the empty sequence, then $(p, \genfrac{}{}{0pt}{}{a}{B}, \genfrac{}{}{0pt}{}{S}{L}) \in \delta'(q, \genfrac{}{}{0pt}{}{a}{X})$,

- point 2f changes to: if $\alpha = X_1X_2 \ldots X_r$, then $\delta'(p_1^t, \genfrac{}{}{0pt}{}{a}{B}) = \{(p, \genfrac{}{}{0pt}{}{a}{B}, \genfrac{}{}{0pt}{}{S}{L})\}$.

In transitions shown above, if $p \in F$, then it should be replaced with q_A and if $p \in R$, then it should be replaced with q_R.

The above design of the Turing machine simulating a pushdown automaton shows that the Turing machine has the stop property and it accepts input if and only if the pushdown automaton accepts the same input. □

6.5 Problems

Problem 6.1. Design a pushdown automaton accepting the language:

$$L = \{w \in \{a,b\}^* : \#_a w = 2 * \#_b w\}$$

Solution. A deterministic pushdown automaton solves the problem:

$$M = (\{p_1, p_2, ACC, REJ\}, \{a,b\}, \{A, B, \rhd\}, \delta, p_1, \rhd, \lhd, \{ACC\}, \{REJ\})$$

where the transition function $\delta : Q \times \Sigma \times \Gamma \to Q \times \Gamma^*$ is given in Table 6.3.

Table 6.3: The transition function of the pushdown automaton designed in Problem 6.1.

$\delta(p_0)$	a	b	$\lhd$
A	(p_1, A)	(p_0, ε)	REJ
B	(p_1, B)	(p_0, BB)	REJ
$\rhd$	$(p_1, \rhd)$	$(p_0, B\rhd)$	ACC

$\delta(p_1)$	a	b	$\lhd$
A	(p_0, AA)	(p_1, ε)	REJ
B	(p_0, ε)	(p_1, BB)	REJ
$\rhd$	$(p_0, A\rhd)$	$(p_1, B\rhd)$	REJ

A brief description of an algorithm:
- the number of pairs of letters a is compared to the number of letters b in an input word:
  - the state p_0 stands for an even number of as already read from the input. Next, odd, a read from the input causes switch to the state p_1 and does not change the stack,
  - the state p_1 stands for an odd number of as already read from the input. Next, even a read from the input switches the state to p_0 and causes either deleting B from the stack or pushing A on the stack. Deleting B from the stack corresponds to matched pair of two as and one b while pushing A on the stack means that two as read from the input does not meet the corresponding b,

  - due to assumptions of the above two points, in state p_1, each pair of a and B or pair of b and A stands for a matched pair of two as and one b; otherwise, A or B corresponding to a or b is pushed on the stack,
  - alike, in state p_0, the pair of b and A stands for a matched pair of two as and one b,
  - stack symbols A and B on the stack correspond to superfluous pairs of letters a or letters b,
- an automaton accepts by the accepting state ACC, though the transition to the accepting state is made only if the stack is empty.

Examples of computation of the above automaton for input words *abaaaabab* and *aaaaab*:

$$\begin{aligned}\triangleright p_0\, a\,b\,a\,a\,a\,a\,b\,a\,b \triangleleft &\succ \triangleright p_1\, b\,a\,a\,a\,a\,b\,a\,b \triangleleft \succ \triangleright B\, p_1\, a\,a\,a\,a\,b\,a\,b \triangleleft \\ &\succ \triangleright p_0\, a\,a\,a\,b\,a\,b \triangleleft \succ \triangleright p_1\, a\,a\,b\,a\,b \triangleleft \succ \triangleright A\, p_0\, a\,b\,a\,b \triangleleft \\ &\succ \triangleright A\, p_1\, b\,a\,b \triangleleft \succ \triangleright p_1\, a\,b \triangleleft \succ \triangleright A\, p_0\, b \triangleleft \succ \triangleright p_0 \triangleleft \succ \triangleright A\,C\,C \triangleleft \\ \triangleright p_0\, a\,a\,a\,a\,a\,b \triangleleft &\succ \triangleright p_1\, a\,a\,a\,a\,b \triangleleft \succ \triangleright A\, p_0\, a\,a\,a\,b \triangleleft \succ \triangleright A\, p_1\, a\,a\,b \triangleleft \\ &\succ \triangleright A\,A\, p_0\, a\,b \triangleleft \succ \triangleright A\,A\, p_1\, b \triangleleft \succ \triangleright A\, p_1 \triangleleft \succ \triangleright R\,E\,J \triangleleft\end{aligned}$$

Problem 6.2. Design a pushdown automaton accepting the language $L = \{ww^R : w \in \{a,b\}^*\}^*$, that is, the language of sequences of even length palindromes over the alphabet $\{a,b\}$.

Solution. The following nondeterministic pushdown automaton solves the problem:

$$M = (\{p_1, p_2, ACC, REJ\}, \{a,b\}, \{A, B, \triangleright\}, \delta, p_1, \triangleright, \triangleleft, \{ACC\}, \{REJ\})$$

where the transition function $\delta : Q \times (\Sigma \cup \{\varepsilon\}) \times \Gamma \to Q \times \Gamma^*$ is given in Table 6.4.

Table 6.4: The transition function of the pushdown automaton designed in Problem 6.2.

$\delta(p_1)$	a	b	ε	$\triangleleft$
A	(p_1, AA)	(p_1, BA)	(p_2, A)	REJ
B	(p_1, AB)	(p_1, BB)	(p_2, B)	REJ
$\triangleright$	$(p_1, A\triangleright)$	$(p_1, B\triangleright)$	REJ	ACC

$\delta(p_2)$	a	b	ε	$\triangleleft$
A	(p_2, ε)	REJ	REJ	REJ
B	REJ	(p_2, ε)	REJ	REJ
$\triangleright$	REJ	REJ	$(p_1, \triangleright)$	REJ

A brief description of the algorithm:

- a part of an input word is checked if it is a palindrome of even length:
  - in the state, p_1 stack symbols A and B corresponding to letters a and letters b are pushed on the stack for the first half of the part of an input word,
  - a nondeterministic transition is made to the state p_2 to compare the contents of the stack (the first half of the palindrome) to incoming input symbols,
  - if the input is nonempty, the (deterministic) transition is made to the state p_1 in order to continue checking the next part of the input,
- an automaton accepts by the accepting state ACC, though the transition to the accepting state is made from the state p_1 only if the stack is empty. Note that the empty word is also accepted by the transition from the state p_1.

A computation for the word $w = aaaa$ is shown in Figure 6.5. Notice that the automaton found two different splits of the input word to palindromes: the first one takes the whole input as palindrome and the second split divides the input word into two halves being palindromes as well.

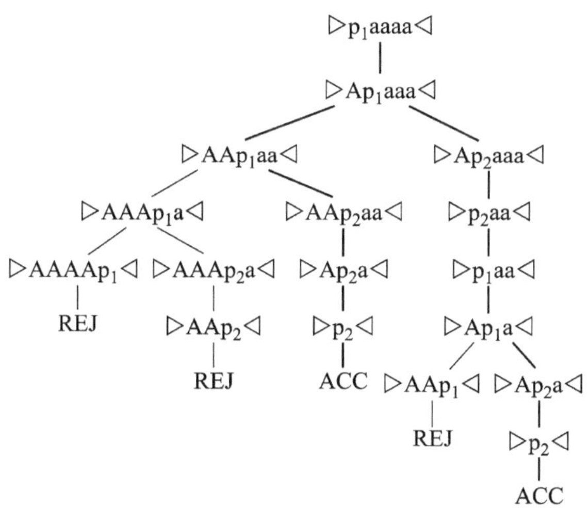

Figure 6.5: The computation of pushdown automaton designed in Problem 6.2 for the input word w = *aaaa*.

Problem 6.3. Design a pushdown automaton accepting the language:

$$L = \{w = a_1\, a_2 \ldots a_n \in \{a, b\}^* : \#_a\, w = \#_b\, w \text{ and } a_1 = a_n\}$$

that is, language of words over the alphabet $\{a, b\}$ with the same number of letters a and b and with the first and the last letters equal.

Hint: design a deterministic pushdown automaton. Store in states of an automaton the first and the recently read input letters.

Problem 6.4. Design a pushdown automaton accepting the language:

$$L = \{w = a_1 a_2 \ldots a_n \in \{a,b\}^* : \#_a w = \#_b w \wedge (\forall u, uv = w)\#_a u \geq \#_b u\}$$

that is, language of words over the alphabet $\{a,b\}$ which have the same number of a's and b's and have no more b's than a's in all its prefixes.

Hint: design a deterministic pushdown automaton which:
- pushes A on the stack for a on the input,
- pops A from the stack for B on the input,
- rejects the input only for empty stack and b on the input.

Problem 6.5. Design a pushdown automaton accepting the language:

$$L = \{w = \in \{a,b\}^* : w = a^n b^n,\ n > 0\}^*$$

that is, the language of words over the alphabet $\{a,b\}$ which is a concatenation of an arbitrary number of equal length sequences of a's and b's.

Hint: design a deterministic pushdown automaton.

Problem 6.6. Let $L_1 = L(M_1)$ and $L_2 = L(M_2)$ are languages accepted by pushdown automata M_1 and M_2. Build a pushdown automaton M, which accepts:
a) union of languages L_1 and L_2, that is, the language $L = L_1 \cup L_2$,
b) concatenation $L = L_1 \circ L_2$ of languages L_1 and L_2,
c) Kleene closure of the language L_1, that is, the language $L = (L_1)^*$.

Hints: design nondeterministic pushdown automata, which respectively:
a) an automaton makes nondeterministic ε-move choosing either M_1, or M_2 and than applies computation of the chosen automaton,
b) an automaton nondeterministically splits the input word into two parts: it follows computation of M_1 for a part of the input word, which could be accepted by M_1, and then switches to the M_2 and then follows its computation for remaining part of the input word,
c) an automaton nondeterministically splits the input word into several parts: it follows computation of M for a part of the input word, which M could accept. It then begins new computation of M for the remaining part of the input word with the possible beginning of new computation of M.

7 Finite automata

Finite automata are the simplest model of computation. They can be derived from pushdown automata by removing the stack. Finite automata are finite structures without any potentially infinite parts as, for instance, a stack in pushdown automata or a tape in Turing machines. Despite these limitations, finite automata are important theoretical and practical tools. Three classes of finite automata are distinguished: deterministic, nondeterministic and those with ε-transitions.

7.1 Deterministic finite automata

Deterministic model of finite automata is the simplest one among three types: deterministic, nondeterministic and with ε-transitions. It is the simplest one in terms of a description of automata as well as of computation realized for a given input word. From now on the term *finite automata* will denote *deterministic* finite automata. Any reference to nondeterministic finite automata or finite automata *with ε-transitions* will be explicitly acknowledged.

Definition 7.1. A deterministic finite automaton is a system

$$M = (Q, \Sigma, \delta, q_0, F)$$

with the following components:

Q a finite set of states,
Σ a finite input alphabet,
q_0 the initial state, $q_0 \in Q$,
F a set of accepting states, $F \subset Q$,
δ a transition function, $\delta : Q \times \Sigma \to Q$.

A transition function of a finite automaton is a total function, that is, it is defined for all its pairs of arguments (a transition function of Turing machines and pushdown automata could be undefined for some arguments).

A finite automaton could be interpreted as a structure shown in Figure 7.1. It consists of:

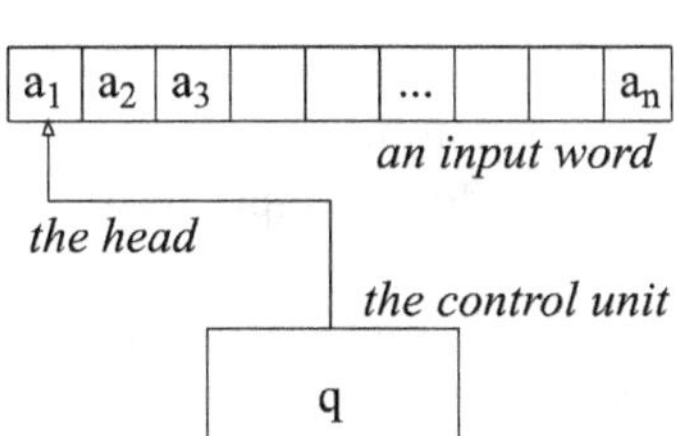

Figure 7.1: A deterministic finite automaton.

https://doi.org/10.1515/9783112223123-009

- a control unit, it is in a state of $q \in Q$,
- an input tape holding input data $a_1a_2 \dots a_n$,
- an input head, which reads an input symbol and shifts right.

Like in the case of pushdown automata, a finite automaton is aimed at accepting a language, that is, answering if an input word belongs to the language or does not. Computation of a given finite automaton is done according to the following intuitive procedure:

1. the initial configuration of a given finite automaton is described as follows:
   a. an input data, a word $w = a_1a_2 \dots a_n$ over input alphabet Σ, is stored on the input tape, cf. Figure 7.1,
   b. the head of the input tape is placed over the first (leftmost) symbol of the input word,
   c. the control unit is in the initial state q_0,
2. based on:
   a. a state q of the control unit,
   b. an input symbol a,

   the automaton realizes the following actions:
   c. a state p of the transition function $\delta(q, a)$ is computed, where $a \in \Sigma$ is the input symbol and $q, p \in Q$ are states,
   d. the input head is shifted right,
   e. the control unit switches to the state p,
3. if the input is not empty, then computation goes to point 2; otherwise, computation is terminated and an automaton responds to a state of the control unit.

Of course, the computation of a finite automaton always terminates.

Note that finite automata, like pushdown automata, cannot store output information, so they can only accept languages and cannot compute functions or solve problems.

Now we define a configuration, a transition relation and a computation of finite automata.

Definition 7.2. A configuration of a finite automaton $M = (Q, \Sigma, \delta, q_0, F)$ is the following sequence of symbols:

$$q\ \omega$$

where:

- $q \in Q$ is the current state of the control unit of a finite automaton,
- ω is the current input, the first symbol of ω is the symbol under the input head.

Note that a sequence of symbols ω is a word over the input alphabet Σ. It is empty after the termination of a computation. It can never be infinite.

For instance, the initial configuration (configuration) is "$q_0\, a_1\, a_2\, \ldots\, a_n$", where q_0 is the initial state, $w = a_1\, a_2\, \ldots\, a_n$ is the input data. On the other hand, a configuration "q" informs that the control unit of a finite automaton is in a state q and an input symbols have already been read.

As in the case of Turing machines and pushdown automata, we will use the symbol $\succ$ to denote a transition of a finite automaton. The transition symbol $\succ$ may be supplemented with a finite automaton name $\succ_M$ to emphasize that a transition concerns a given finite automaton. It also can be supplemented with a superscript $\succ^k$ to notify k transitions have been completed.

If $q\ a_i\, a_{i+1}\, \ldots\, a_n$ is a configuration and $\delta(q, a_i) = p$ is a transition function applicable for this configuration, then a next step description after a transition is made is $p\ a_{i+1}\, \ldots\, a_n$. These two step descriptions create a transition of a finite automaton, which is denoted as:

$$q\ a_i\ a_{i+1}\ \ldots\ a_n \succ p\ a_{i+1}\ \ldots\ a_n$$

Definition 7.3. Transitions of a finite automaton create a binary relation in the space of all possible configurations of the automaton, that is, any two configurations are related if and only if the latter is derived from the former one by application of a transition function. This relation is called the transition relation of a given finite automaton. The transitive closure of the transition relation is denoted by $\succ^*$.

Definition 7.4. A computation of a finite automaton $M = (Q, \Sigma, \delta, q_0, F)$ is a sequence of configurations $\eta_1, \eta_2, \ldots, \eta_n$ such that η_1 is the initial configuration, η_T is the final configuration and a pair of any two successive configurations belongs to the transition relation. A computation is denoted as $\eta_1 \succ \eta_2 \succ \cdots \succ \eta_n$.

Remark 7.1. A computation of a finite automaton is a finite sequence of configurations

$$q_0\, a_1\, a_2\, a_3\, \ldots\, a_n \succ q_{i_1}\, a_2\, a_3\, \ldots\, a_n \succ q_{i_2}\, a_3\, \ldots\, a_n \succ q_{i_{n-1}}\, a_n \succ q_{i_n}$$

Because any transition consists of reading an input symbol and switching to a state, computation will be shown in a simpler form:

$$q_0\ a_1\ q_{i_1}\ a_2\ q_{i_2}\ a_3\ \ldots\ a_n\ q_{i_{n-1}}\ a_n\ q_{i_n}$$

where only input symbols under the head are displayed.

Now we give a formal definition of acceptance of an input by a finite automaton.

Definition 7.5. A finite automaton accepts its input if and only if the pair of the initial configuration and a final configuration belongs to the transitive closure of a transitive relation, that is, $\eta_1 \succ^* \eta_T$, where η_1 is an initial configuration and η_T is a final configuration, that is, the state in η_T is an accepting one.

Based on the above discussion, we now give formal definitions of some concepts.

Definition 7.6. The language accepted by a finite automaton $M = (Q, \Sigma, \delta, q_0, F)$ is the set of words $w \in \Sigma^*$ accepted by a finite automaton.

Example 7.1. Design a finite automaton accepting binary numbers without non-significant zeros. Note that binary numbers without non-significant zeros are zero and numbers, which begin with 1:

$$L = \{0, 1, 10, 11, 100, 101, 110, 111, 1000, 1001, 1010, \ldots\}$$

Solution. The following automaton accepts binary words without non-significant zeros:

$$M = (Q = \{q_0, q_1, q_2, q_3\}, \Sigma = \{0, 1\}, \delta, q_0, F = \{q_1, q_3\})$$

where the transition function $\delta : Q \times \Sigma \to Q$ is given in Table 7.1. In this table, the initial state is marked by an arrow left of the initial state and accepting states are marked with arrows right of them.

Table 7.1: The transition function of the finite automaton designed in Example 7.1.

δ	0	1
$\to q_0$	q_1	q_3
$q_1 \to$	q_2	q_2
q_2	q_2	q_2
$q_3 \to$	q_3	q_3

A finite automaton can also be shown in the form of a transition diagram. For instance, a transition diagram of the above automaton is illustrated in Figure 7.2. The initial state is indicated by the arrow going to it, accepting states are double circled and transitions are shown as labeled arrows between states.

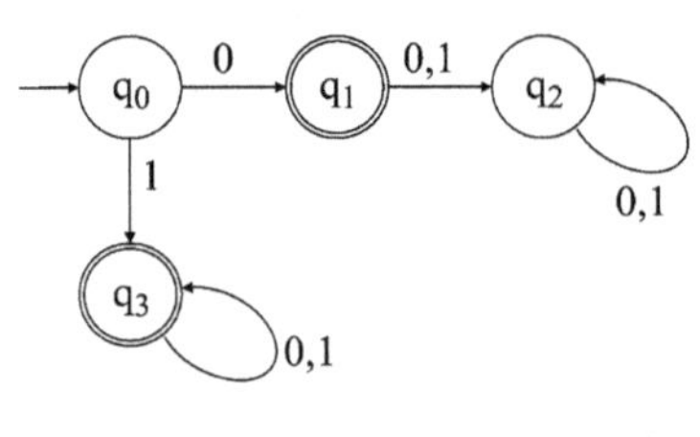

Figure 7.2: Transition diagram of the finite automaton designed in Example 7.1.

It is clear that this automaton accepts the language L:
- computation for any binary word, which begins with 1 (it is a binary number without non-significant zeros), goes to the state q_3 and stays in this state, waiting for any

more input binary digits until reaching termination of the computation. Because q_3 is an accepting state, then the word is accepted,
- computation for 0 (0 is a binary number without non-significant zeros) goes and terminates in q_1, which is an accepting state. Thus, 0 is accepted,
- any other binary word begins with 0 and has more than one binary digit. Such words are binary numbers with non-significant zeros. For such a word, computation goes to q_1 and then to q_2 and then stays in q_2 for any more input binary digits until termination. Such a word will not be accepted because q_2 is not an accepting state.

Examples of computation:
- q_0 – the word ε is not accepted because computation for this word is terminated in a state q_0, which is not an accepting one. Note that the empty word ε is considered not to be a number,
- $q_0\ 1\ q_3\ 1\ q_3\ 0\ q_3\ 1\ q_3\ 0\ q_3$ – the word 11010 is accepted because computation for this word is terminated in the accepting state q_3,
- $q_0\ 0\ q_1$ – the word 0 is accepted because computation for this word is terminated in the accepting state q_1,
- $q_0\ 0\ q_1\ 1\ q_2\ 0\ q_2\ 1\ q_2\ 0\ q_2$ – the word 01010 is not accepted because computation for this word is terminated in the state q_3, which is a not accepting one.

Example 7.2. Design a finite automaton, which accepts the language L of binary words with sequences of the same letter not longer than 3, for instance: $\varepsilon \in L$, $1000111 \in L$, $10000111 \notin L$.

Solution. The following finite automaton accepts the language:

$$M = (Q, \Sigma, \delta, q_0, F)$$

where:
- $Q = \{q_0, q_{01}, q_{02}, q_{03}, q_{11}, q_{12}, q_{13}, q_R\}$ – the set of states,
- $\Sigma = \{0, 1\}$ – the input alphabet,
- $F = \{q_0, q_{01}, q_{02}, q_{03}, q_{11}, q_{12}, q_{13}\}$ – the set of accepting states,
- δ – the transition function given in Table 7.2. A transition diagram of this automaton is shown in Figure 7.3.

The finite automaton counts successive 0's and successive 1's. Successive 0's are counted by moving to states q_{01}, q_{02}, q_{03}. Likewise, successive 1's are counted in states q_{11}, q_{12}, q_{13}. Words with more than three the same successive letters are not included in the language. For that reason, when the fourth 0 or fourth 1s appears, that is, the control unit is in the state q_{03} or q_{13}, then the transition is made to the state q_R, which is not an accepting state and, in this way, it is a trap for words with successive four or more 0s or 1s.

Examples of computation:
- q_0 – the word ε is accepted because computation for this word is terminated in the state q_0, which is an accepting one,

Table 7.2: The transition function of the finite automaton designed in Example 7.2.

δ	0	1
$\rightarrow q_0 \rightarrow$	q_{01}	q_{11}
$q_{01} \rightarrow$	q_{02}	q_{11}
$q_{02} \rightarrow$	q_{03}	q_{11}
$q_{03} \rightarrow$	q_R	q_{11}
$q_{11} \rightarrow$	q_{01}	q_{12}
$q_{12} \rightarrow$	q_{01}	q_{13}
q_{13}	q_{01}	q_R
q_R	q_R	q_R

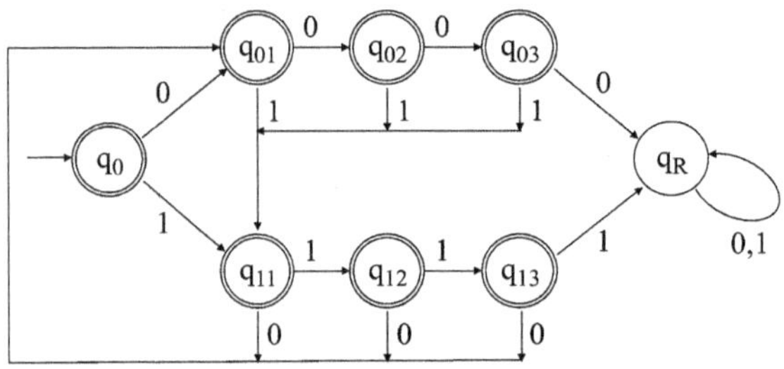

Figure 7.3: Transition diagram of the finite automaton accepting the language given in Example 7.2.

- q_0 1 q_{11} 1 q_{12} 0 q_{01} 0 q_{02} 0 q_{03} 1 q_{11} – the word 110001 is accepted because computation for this word is terminated in the state q_{11}, which is an accepting one,
- q_0 1 q_{11} 1 q_{12} 0 q_{01} 0 q_{02} 0 q_{03} 0 q_R 1 q_R – the word 1100001 is not accepted because computation for this word is terminated in the state q_R, which is not an accepting one.

A transition relation identified in Definition 7.3 yields exactly one state for a given configuration of a finite automaton. Namely, for a given state q and a given symbol of an input alphabet a, there is exactly one state p related to the given state q. The state p is yielded by a transition function: $p = \delta(q, a)$. This property comes from the definition of a transition function, which is total and its value is a state: $\delta : Q \times \Sigma \rightarrow Q$. In other words, for a given state, exactly one state is related to it with regard to a given symbol of an input alphabet. The transitive closure of a transition relation is a function as well, that is, for a given state, exactly one state is related with regard to a given sequence of symbols of an input alphabet. This property is exploited in a definition of a so-called *closure of transition function.*

Definition 7.7. A closure of a transitions function δ of a given deterministic finite automaton $M = (Q, \Sigma, \delta, q_0, F)$ is the function $\hat{\delta}$ satisfying the following conditions

1. $\hat{\delta} : Q \times \Sigma^* \rightarrow Q$,
2. $(\forall q \in Q)\, \hat{\delta}(q, \varepsilon) = q$,
3. $(\forall q \in Q)(\forall a \in \Sigma)(\forall w \in \Sigma^*)\hat{\delta}(q, wa) = \delta(\hat{\delta}(q, w), a)$.

For deterministic finite automata, the closure of a transition function extends its domain from an input alphabet of an automaton to the set of all words over an input alphabet. The closure of a transition function applied to an initial configuration of a finite automaton immediately yields a result of computation $\hat{\delta}(q_0, w)$ of this automaton for the initial state q_0 and an input word $w = a_1a_2 \dots a_n$.

Remark 7.2. The restriction of the closure $\hat{\delta}$ of a transition function δ to the domain of δ is equal to δ:

$$\hat{\delta}|_{Q\times\Sigma} = \delta$$

For that reason, both a transition function (of a deterministic finite automaton) and its closure are denoted by the same symbol δ unless it might result in some misinterpretation.

7.2 Nondeterministic finite automata

In the previous section, we have studied deterministic finite automata. In this section, we introduce and discuss nondeterministic finite automata. Nondeterministic finite automata, compared to their deterministic counterparts, differ in the form of the transition function. A transition function of nondeterministic finite automata allows for choosing a transition among several states. This new feature simplifies solutions of problems, though it does not increase the computational abilities of finite automata. Below, we will prove that both classes of automata, that is, deterministic finite automata and nondeterministic finite automata, are equivalent with regard to the nature of the accepted languages. This means that each of these two classes of finite automata accepts the same class of languages. In other words, for a given automaton of one class, we can build an equivalent automaton of another class. The proof of equivalence of both classes of automata is a constructive one, which means that it formulates a method of designing a deterministic finite automaton, which is equivalent to a given nondeterministic one. Of course, a trivial design of nondeterministic automata equivalent to deterministic ones is also mentioned.

Definition 7.8. A nondeterministic finite automaton is a system

$$M = (Q, \Sigma, \delta, q_0, F)$$

where:
- δ – a transition function, $\delta : Q \times \Sigma \to 2^Q$,
- Q, Σ, q_0, F are the same as given in Definition 7.1.

A transition function of nondeterministic finite automata is a total function, the same as for deterministic finite automata, that is, it is defined for all pairs of arguments.

The values of a transition function are subsets of a set of states, including the empty set (which is a subset of the set of all states).

Interpretation of nondeterministic finite automata is similar to the interpretation of deterministic ones, that is, as it is shown in Figure 7.1.

A definition of a configuration (step description) of nondeterministic finite automata is identical with the definition of a configuration of deterministic finite automata, cf. Definition 7.2.

Remark 7.3. Like in the case of deterministic finite automata, nondeterministic ones are aimed at accepting a language. The computation of a given nondeterministic finite automaton is done according to the following intuitive procedure:

1. the initial configuration of a given finite automaton is described as follows:
   a. an input data, a word $w = a_1 a_2 \dots a_n$ over input alphabet Σ, is stored on the input tape, cf. Figure 7.1,
   b. the head of the input tape is placed over the first (leftmost) symbol of the input word,
   c. the control unit is in the initial state q_0,
2. based on:
   a. a state q of the control unit,
   b. an input symbol a,

   the automaton realizes the following actions:

   c. a set of states $\{p_1, p_2, \dots, p_k\}$ being the value of the transition function $\delta(q, a)$ is computed, where $a \in \Sigma$ is the input symbol and $q, p_1, p_2, \dots, p_k \in Q$ are states,
   d. if the set yielded by a transition function is empty, then the computation is terminated and input is rejected, otherwise,
   e. a state p_i is nondeterministically picked up from the set of states $\{p_1, p_2, \dots, p_k\}$,
   f. the input head is shifted right,
   g. the control unit switches to the state p_i,
3. if the input is not empty, then computation goes to point 2; otherwise, computation is terminated and an automaton responds with a state of the control unit.

Of course, the computation of a nondeterministic finite automaton always terminates, as we have seen it for deterministic finite automata.

The symbols $\succ$, $\succ_M$ and $\succ^k$ are used in the usual way, refer to our discussion on deterministic finite automata.

Definition 7.9. Transitions of a nondeterministic finite automaton form a binary relation in the space of all possible configurations of the automaton, that is, any two configurations are related if and only if the later one is derived from the former one by a transition in terms of point 2. of Remark 7.3. This relation is called the transition relation of a given nondeterministic finite automaton. The transitive closure of the transition relation is denoted by $\succ^*$.

Definition 7.10. An interpretation of a computation of a nondeterministic finite automaton

$$M = (Q, \Sigma, \delta, q_0, F)$$

is interpreted as a tree such that:
- its nodes are labeled by configurations of the automaton,
- its root is labeled by the initial configuration,
- for any node η_p, its every child ηc is related to it in the transition relation, that is, $\eta_p \succ \eta_c$.

Remark 7.4. Computation of a nondeterministic automaton is a sequence of configurations, similar in the case of a deterministic one. Consecutive configurations are set by nondeterministic choice of a state from the value of the transition function, as it was discussed in the case of nondeterministic Turing machine, cf. Definition 5.18 and Remark 5.4. However, since computation is usually interpreted as a tree, we will call the computation tree a computation.

Remark 7.5. Note that a computation tree of a nondeterministic finite automaton is a k-tree, where k is the maximal number of values yielded by the transition function for given arguments, that is, k is the degree of nondeterminism. Moreover, the edges of every level of such a tree are labeled with the same input symbol.

Remark 7.6. A nondeterministic finite automaton accepts its input if and only if the computation tree has a path from the root to a leaf labeled by an accepting configuration.

Remark 7.7. Formal definitions of acceptance of input and of a language accepted by a nondeterministic finite automaton are identical with the respective definitions of deterministic finite automata, cf. Definition 7.4 and Definition 7.6.

Example 7.3. Design a finite automaton accepting the language L of binary words with three successive 0's or three successive 1's.

Solution. The following automaton accepts the language L:

$$M = (Q = \{q_0, q_{10}, q_{20}, q_{11}, q_{21}, q_A\}, \Sigma = \{0, 1\}, \delta, q_0, F = \{q_A\})$$

where the transition function $\delta : Q \times \Sigma \to 2^Q$ is given in Table 7.3. A transition diagram of this automaton is shown in Figure 7.4.

An example of computation of the above nondeterministic automaton is given in Figure 7.5. The input word $w = 010001$ is accepted because the computation tree has a path from the root to an accepting leaf.

An algorithm realized by the automaton is shown below:
- the automaton stays in the state q_0 reading input symbols until a sequence of three successive 0s or three successive 1s arrives,

Table 7.3: The transition function of the finite automaton designed in Example 7.3.

δ	0	1
$\to q_0$	$\{q_0, q_{10}\}$	$\{q_0, q_{11}\}$
q_{10}	$\{q_{20}\}$	$\emptyset$
q_{20}	$\{q_A\}$	$\emptyset$
q_{11}	$\emptyset$	$\{q_{21}\}$
q_{21}	$\emptyset$	$\{q_A\}$
$q_A \to$	$\{q_A\}$	$\{q_A\}$

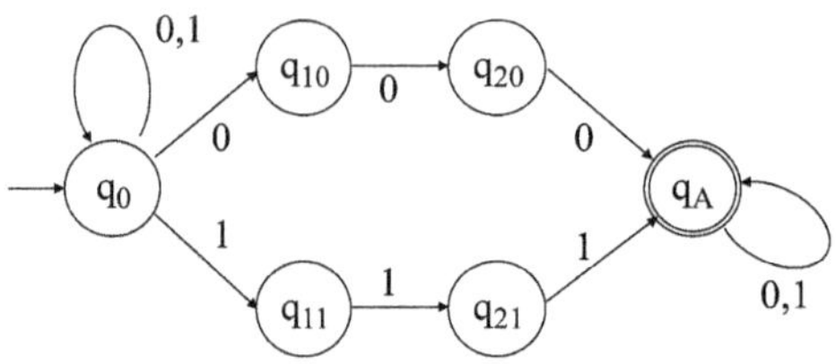

Figure 7.4: Transition diagram of the finite automaton designed in Example 7.3.

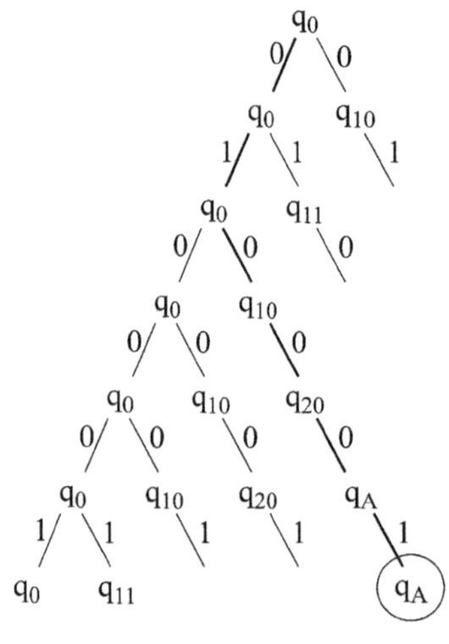

Figure 7.5: The computation of the automaton designed in Example 7.3 for the input word $w = 010001$.

- when the beginning of a sequence of three successive 0s or 1s is nondeterministically encountered, a transition is made to the state q_{10} or q_{11}, respectively,
- next two consecutive 0s or 1s are counted by transitions to states q_{20} and q_A or to states q_{21} and q_A,
- when the state q_A has been reached, computation stays in this state for next coming input symbols,
- input words, for which the accepting state q_A is reached, are accepted and only such words.

The transition relation of nondeterministic automata yields a subset of a set of states for a given configuration (unlike the transition relation of deterministic automata, which yields exactly a single state). For a given state q and for a given symbol a of an input alphabet, states $p \in P \subset Q$, where P is the value of the transition function, are related to the given state q. This property comes from the definition of a transition function, which is total and its values are subsets of the set of states: $\delta : Q \times \Sigma \to 2^Q$. The transitive closure of a transition relation also relates subsets of the set Q for a given state and a given symbol of an input alphabet. The following definition provides details of a so-called *closure of transition function* for nondeterministic automata.

Definition 7.11. A closure of a transitions function δ of a given nondeterministic finite automaton $M = (Q, \Sigma, \delta, q_0, F)$ is the function:
1. $\hat{\delta} : Q \times \Sigma^* \to 2^Q$
2. $(\forall q \in Q)\, \hat{\delta}(q, \varepsilon) = \{q\}$
3. $(\forall q \in Q)(\forall a \in \Sigma)(\forall w \in \Sigma^*)\hat{\delta}(q, wa) = \delta(\hat{\delta}(q, w), a)$

where the following designation stands for $\delta(\hat{\delta}(q, w), a)$: $\delta(P, a) = \bigcup_{p \in P} \delta(p, a)$, for any $P \subset Q$.

As is in case of deterministic finite automata, the closure of a transition function of a nondeterministic finite automaton extends a domain of a transition function from an input alphabet of an automaton to the set of all words over an input alphabet. The closure of a transition function applied to an initial configuration of a finite automaton immediately yields the result of computation $\hat{\delta}(q_0, w)$ of this automaton for the initial state q_0 and an input word $w = a_1 a_2 \ldots a_n$.

Remark 7.8. For nondeterministic finite automata, the restriction of the closure $\hat{\delta}$ of a transition δ to the domain $Q \times \Sigma$ is equal to the transition function δ:

$$\hat{\delta}|_{Q \times \Sigma} = \delta$$

For that reason, both a transition function of a nondeterministic finite automaton and its closure are denoted by the same symbol δ unless it may cause misinterpretation.

Remark 7.9. An input word w is accepted by the nondeterministic finite automaton $M = (Q, \Sigma, \delta, q_0, F)$ if and only if $\delta(q_0, w) \cap F \neq \varnothing$.

A question is raised whether the class of nondeterministic finite automata is equivalent to the class of deterministic ones. The answer is positive.

On the one hand, a deterministic automaton is a nondeterministic one with the degree of nondeterminism equal to one, that is, which does not exploit nondeterminism. Formally, a deterministic automaton can be turned into a nondeterministic one by replacing a value of its transition function, which is a state, by the set including only this single state.

Conversely, we will prove that, for a given nondeterministic automaton, there exists an equivalent deterministic one, that is, accepting the same language. An idea of designing such a deterministic automaton is based on observation of computations of both classes of automata. The idea relies on turning a computation of a nondeterministic automaton, which is a tree of configurations, into a sequence of configurations: states of every level of a computation tree are collected into a set of them. In such a way, a tree is twisted into a sequence of alternating sets of states and symbols of the input alphabet. Such a sequence corresponds to the computation of a deterministic finite automaton, supposing that sets of states represent potential states of a deterministic finite automaton.

The details behind these intuitive observations are outlined as follows:

Proposition 7.1. *The class of nondeterministic finite automata is equivalent to the class of deterministic ones.*

Proof. First, the following deterministic automaton:

$$M = (Q, \Sigma, \delta, q_0, F)$$

is equivalent to a nondeterministic one:

$$M = (Q, \Sigma, \delta', q_0, F)$$

such that $\delta' : Q \times \Sigma \to 2^Q$, $\delta'(q, a) = \{\delta(q, a)\}$ for $q \in Q$ and $a \in \Sigma$.

Second, let us assume that now a nondeterministic finite automaton is:

$$M = (Q, \Sigma, \delta, q_0, F)$$

An equivalent deterministic automaton is denoted as follows:

$$M' = (Q', \Sigma, \delta', q'_0, F')$$

where:
- $Q' \cong 2^Q$ – states of the deterministic automaton correspond to sets of the nondeterministic one. A state of Q' corresponding to a set of states $\{q_{i_1}, q_{i_2}, \ldots, q_{i_j}\}$ will be denoted $[q_{i_1}, q_{i_2}, \ldots, q_{i_j}]$ just to distinguish subsets of Q from states of Q',
- $q'_0 = [q_0]$ – the initial state of M' is the set including the initial set of M,
- $F' = \{[q_{i_1}, q_{i_2}, \ldots, q_{i_j}] \in Q' : \{q_{i_1}, q_{i_2}, \ldots, q_{i_j}\} \cap F \neq \varnothing\}$ – accepting states of M' are labeled by sets of states including an accepting state(s) of M,
- $\delta'([q_{i_1}, q_{i_2}, \ldots, q_{i_j}], a) = [p_{i_1}, p_{i_2}, \ldots p_{i_k}] \Leftrightarrow \bigcup_{l=1}^{j} \delta'(q_{i_l}, a) = \{p_{i_1}, p_{i_2}, \ldots, p_{i_k}\}$ – the transition function of M' takes union of its values (which are sets of states of M), for states included in its argument (which corresponds to a set of states of M).

A formal proof of equivalence of M and M' is based on mathematical induction concerning the length of input word of both automata.

Let us prove that the closure of transition functions of both automata hold the equivalence for any $w \in \Sigma^*$ (note that the same symbol denotes both a transition function and its closure):

$$\delta'([q_0], w) = [q_{i_1}, q_{i_2}, \ldots, q_{i_j}] \Leftrightarrow \delta(q_0, w) = \{q_{i_1}, q_{i_2}, \ldots, q_{i_j}\} \qquad (*)$$

1. for the word of length 0, that is, for the empty word, the equivalence $(*)$: $\delta'([q_0], \varepsilon) = [q_0]$ and $\delta(q_0) = \{q_0\}$ is derived directly from definitions of the closure of transition functions,
2. let us check if the equivalence $(*)$ holds for a word $a_1\, a_1\, \ldots\, a_n\, a = w\, a$. In virtue of the inductive assumption we have that this equivalence holds for any word $w = a_1\, a_1\, \ldots\, a_n \in \Sigma^*$. Let $a \in \Sigma$. Then, based on inductive assumption, we get:
   $\delta'([q_0], wa) = \delta'(\delta'([q_0], w), a) = \delta'([q_{i_1}, q_{i_2}, \ldots, q_{i_j}], a)$ and
   $\delta(q_0, wa) = \delta(\delta(q_0, w), a) = \delta(\{q_{i_1}, q_{i_2}, \ldots, q_{i_j}\}, a)$
   for some set of states $\{q_{i_1}, q_{i_2}, \ldots, q_{i_j}\} \subset Q$.
   Now, let us compute $\delta'([q_{i_1}, q_{i_2}, \ldots, q_{i_j}], a)$ based on the above definition of δ':
   $\delta'([q_{i_1}, q_{i_2}, \ldots, q_{i_j}], a) = [p_{i_1}, q_{i_2}, \ldots, p_{i_k}]$, where $\bigcup_{l=1}^{j} \delta'(q_{i_l}, a) = \{p_{i_1}, p_{i_2}, \ldots, p_{i_k}\}$
   On the other hand
   $\delta(\{q_{i_1}, q_{i_2}, \ldots, q_{i_j}\}, a) = \bigcup_{l=1}^{j} \delta'(q_{i_l}, a) = \{p_{i_1}, p_{i_2}, \ldots, p_{i_k}\}$,
3. utilizing mathematical induction based on 1. and 2. we conclude that the equivalence $(*)$ holds. □

Finally, let us notice that an input word is accepted by the nondeterministic finite automaton M if and only if it is accepted by the deterministic finite automaton M'. This property comes directly from the equivalence $(*)$, Remark 7.9 and definition of the set of accepting states F' of the automaton M'.

Example 7.4. Design a nondeterministic finite automaton accepting the language L of binary words ending with at least three successive 0's or at least three successive 1's.

Solution. The following automaton accepts the language L:

$$M = (Q = \{q_0, q_{10}, q_{20}, q_{30}, q_{11}, q_{21}, q_{31}\}, \Sigma = \{0, 1\}, \delta, q_0, F = \{q_{30}, q_{31}\})$$

where the transition function $\delta : Q \times \Sigma \to 2^Q$ is given in Table 7.4 (also compare Problem 7.1). A transition diagram of this automaton is shown in Figure 7.6. Compare also solution of Problem 7.1.

Let us outline the algorithm realized by the automaton:
- the automaton stays in the state q_0 reading input symbols until a sequence of three successive 0's or three successive 1's arrives and this sequence ends the input word,

Table 7.4: Transition function of the finite automaton designed in Example 7.4.

δ	0	1
$\to q_0$	$\{q_0, q_{10}\}$	$\{q_0, q_{11}\}$
q_{10}	$\{q_{20}\}$	$\emptyset$
q_{20}	$\{q_{30}\}$	$\emptyset$
$q_{30} \to$	$\emptyset$	$\emptyset$
q_{11}	$\emptyset$	$\{q_{21}\}$
q_{21}	$\emptyset$	$\{q_{31}\}$
$q_{31} \to$	$\emptyset$	$\emptyset$

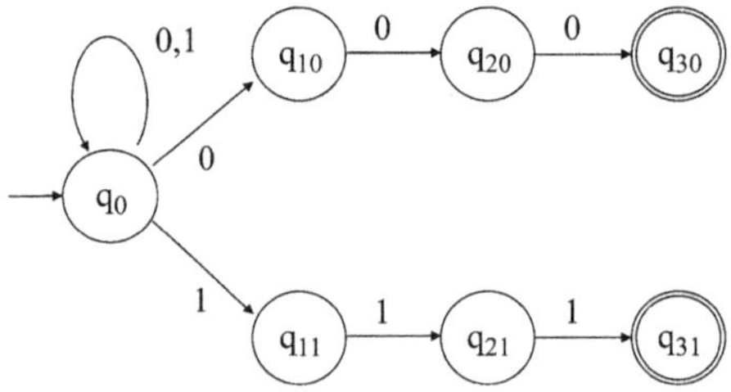

Figure 7.6: Transition diagram of the finite automaton designed in Example 7.4.

- when the beginning of a sequence of successive 0's or 1's is nondeterministically encountered, a transition is made to the state q_{10} or q_{11}, respectively,
- next consecutive 0's or next consecutive 1's are counted by transitions to states q_{20} and q_{30} or to states q_{21} and q_{31},
- only such input words are accepted, for which the accepting state q_{30} or q_{31} is reached.

A deterministic automaton $M' = (Q', \Sigma, \delta, q'_0, F')$ is designed based on Proposition 7.1. According to the designing method, in the automaton M' the set of states Q' has $2^7 = 128$ states and the set of accepting states F' has $2^7 - 2^5 = 96$ states. However, only a few states are reachable from the initial state $q'_0 = [q_0]$. All states, which cannot be obtained in any computation starting from the initial state, are useless from the point of view of acceptance of an input word. For that reason, only states reachable from the initial state may be included in the deterministic finite automaton being designed. The transition table is shown in Table 7.5. The table is filled in by starting with the initial state and then including new states yielded by the transition function δ'. All states, which are not obtained in this way, are useless.

In the transition diagram of this automaton, which is given in Figure 7.7, the useless states (accepting and not accepting) are indicated symbolically. Transitions may go from the useless states to the useful ones, as revealed there in the form of symbolic arrows. However, the transition cannot go from the useful states to the useless ones, so there is no arrow going in the opposite direction.

Table 7.5: Transition function of the deterministic finite automaton equivalent to the one designed in Example 7.4.

δ'	0	1
$\rightarrow [q_0]$	$[q_0, q_{10}]$	$[q_0, q_{11}]$
$[q_0, q_{10}]$	$[q_0, q_{10}, q_{20}]$	$[q_0, q_{11}]$
$[q_0, q_{10}, q_{20}]$	$[q_0, q_{10}, q_{20}, q_{30}]$	$[q_0, q_{11}]$
$[q_0, q_{10}, q_{20}, q_{30}] \rightarrow$	$[q_0, q_{10}, q_{20}, q_{30}]$	$[q_0, q_{11}]$
$[q_0, q_{11}]$	$[q_0, q_{10}]$	$[q_0, q_{11}, q_{21}]$
$[q_0, q_{11}, q_{21}]$	$[q_0, q_{10}]$	$[q_0, q_{11}, q_{21}, q_{31}]$
$[q_0, q_{11}, q_{21}, q_{31}] \rightarrow$	$[q_0, q_{10}]$	$[q_0, q_{11}, q_{21}, q_{31}]$
121 states are useless, their transitions are not shown here		

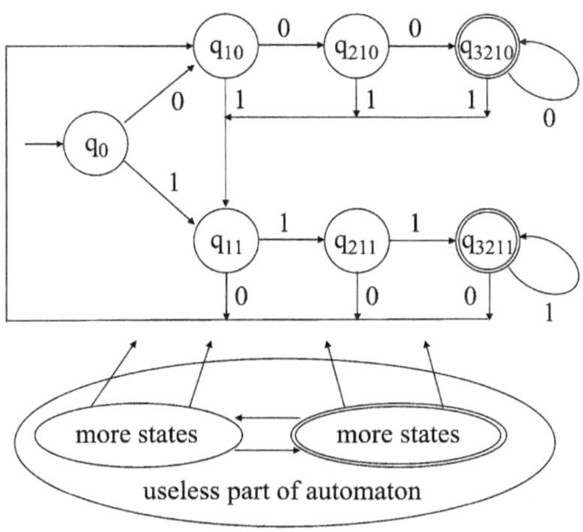

Figure 7.7: Transition diagram of the deterministic finite automaton designed in Example 7.4. Labels of states include indexes only, e. g. q_0 represents the state $[q_0]$, q_{210} represents $[q_0, q_{10}, q_{20}]$, q_{3211} represents $[q_0, q_{11}, q_{21}, q_{31}]$.

As a result of the above notes, useless states will be removed from the deterministic finite automaton designed according to Proposition 7.1. The automaton designed here is:

$$M' = (Q', \Sigma, \delta', q_0', F')$$

where:

- $Q = \{[q_0], [q_0, q_{10}], [q_0, q_{10}, q_{20}], [q_0, q_{10}, q_{20}, q_{30}], [q_0, q_{11}], [q_0, q_{11}, q_{21}], [q_0, q_{11}, q_{21}, q_{31}]\}$,
- δ' – the transition function is given in Table 7.5,
- $q_0 = [q_0]$,
- $F' = \{[q_0, q_{10}, q_{20}, q_{30}], [q_0, q_{11}, q_{21}, q_{31}]\}$.

7.3 Finite automata with ε-transitions

In previous sections of this Chapter, we discussed two classes of finite automata: deterministic finite automata and nondeterministic ones. We proved that both classes are equivalent with regard to accepted languages; that is, for an automaton in one class, an equivalent automaton of another class can be designed. The equivalence of both classes effectively helps in solving problems because we can choose an automaton from a class, which is more appropriate for a problem to be solved. Finite automata with epsilon-transitions, ε-transitions for short, create a third class of finite automata. ε-transitions is the next tool, which may significantly simplify solutions of problems. However, ε-transitions do not increase the computational power of finite automata. In further parts of this section, an equivalence of finite automata with ε-transitions with nondeterministic automata is proved. The proof is a constructive one, that is, it develops the method of designing a nondeterministic finite automaton that is equivalent to a given finite automaton with ε-transitions. A design of a finite automaton with ε-transitions that is equivalent to a given nondeterministic finite automaton is also shown.

Definition 7.12. A finite automaton with ε-transitions is a system

$$M = (Q, \Sigma, \delta, q_0, F)$$

where:
- δ – a transition function, $\delta : Q \times (\Sigma \cup \{\varepsilon\}) \to 2^Q$,
- Q, Σ, q_0, F stay the same as described in Definition 7.1.

A transition function of finite automata with ε-transitions is a total function defined for states and for symbols of the input alphabet with the empty word attached. The type values of a transition function stay the same as for nondeterministic finite automata; that is, they are subsets of a set of states. Thus, finite automata with ε-transitions are nondeterministic ones.

The difference between nondeterministic finite automata and finite automata with ε-transitions lays in the ability of the last-mentioned one to make a transition without checking an input symbol. Let us recall that we already considered a property ε-transitions of pushdown automata. When an ε-transition is made, an input symbol is not checked. This means that ε-transitions are made only for a given state of a current configuration. From another point of view, a configuration for ε-transition is only a state and no input symbol.

A definition of a configuration (step description) of a finite automaton with ε-transitions is identical with the definition of a configuration of deterministic finite automata, cf. Definition 7.2.

An obvious property $\varepsilon \circ \varepsilon = \varepsilon$ leads to an interesting idempotent operation on states for a given finite automaton with ε-transitions

$$M = (Q, \Sigma, \delta, q_0, F)$$

The operation is called epsilon closure, εCl for short. It is defined as follows:

$$\varepsilon Cl(q) = \{p \in Q : p \in \delta^*(q,\varepsilon)\} \quad \text{for } q \in Q$$

where:

$$\delta^* : Q \to 2^Q, \delta^*(q,\varepsilon) = \bigcup_{k=0}^{\infty} \delta^k(q,\varepsilon)$$

and

$$\delta^0(q,\varepsilon) = \{q\}$$
$$\delta^{k+1}(q,\varepsilon) = \delta(\delta^k(q,\varepsilon))$$

where

$$\delta(P,X) = \bigcup_{p\in P} \delta(p,X) \quad \text{for } P \subset Q, X \in (\Sigma \cup \{\varepsilon\})$$

The εCl is idempotent, that is

$$\varepsilon Cl(q) = \varepsilon Cl(\varepsilon Cl(q)) = \bigcup_{p\in \varepsilon Cl(q)} \varepsilon Cl(p)$$

Note that $\delta^*(q,\varepsilon) = \bigcup_{k=0}^{\infty} \delta^k(q,\varepsilon) = \bigcup_{k=0}^{r} \delta^k(q,\varepsilon)$, where r does not need to be greater or equal than the number of states of an automaton. In fact, when a transition diagram of an automaton is considered, then r is equal to the length of the longest path starting at q and having edges labeled with ε.

Example 7.5. Let us consider a finite automaton with ε-transitions:

$$M = (Q, \Sigma = \{a,b\}, \delta, q_0, F = \{q_A\})$$

where:

- $Q = \{q_0, q_{0a}, q_{1a}, q_{2a}, q_{0b}, q_{1b}q_{2b}, q_A\}$,
- $\delta : Q \times (\Sigma \cup \{\varepsilon\}) \to 2^Q$ is the transition function given in Table 7.6. The transition diagram of this automaton is given in Figure 7.8. The automaton accepts words, in which letters come in pairs.

The transition table shown in Table 7.6 has ε closure column. Such a column, which contains ε closure of states, is not an integral part of a transition table. However, it is recommended to attach it for the easiness of further discussion or problem-solving.

Remark 7.10. Like in the case of deterministic and nondeterministic finite automata, finite automata with ε-transitions are aimed at accepting languages. Computation of a given finite automaton with ε-transitions is done according to the following intuitive procedure:

1. the initial configuration of a given automaton is described as follows:
   a. an input data, a word $w = a_1 a_2 \dots a_n$ over input alphabet Σ, is stored on the input tape,

Table 7.6: The transition function of the finite automaton designed in Example 7.5.

δ	a	b	ε	εCl
$\rightarrow q_0$	$\emptyset$	$\emptyset$	$\{q_{0a}, q_{0b}\}$	$\{q_0, q_{0a}, q_{0b}, q_{2a}, q_{2b}, q_A\}$
q_{0a}	$\{q_{1a}\}$	$\emptyset$	$\{q_{2a}\}$	$\{q_0, q_{0a}, q_{0b}, q_{2a}, q_{2b}, q_A\}$
q_{1a}	$\{q_{2a}\}$	$\emptyset$	$\emptyset$	q_{1a}
q_{2a}	$\emptyset$	$\emptyset$	$\{q_A\}$	$\{q_0, q_{0a}, q_{0b}, q_{2a}, q_{2b}, q_A\}$
q_{0b}	$\emptyset$	q_{1b}	$\{q_{2b}\}$	$\{q_0, q_{0a}, q_{0b}, q_{2a}, q_{2b}, q_A\}$
q_{1b}	$\emptyset$	$\{q_{2b}\}$	$\emptyset$	q_{1b}
q_{2b}	$\emptyset$	$\emptyset$	$\{q_A\}$	$\{q_0, q_{0a}, q_{0b}, q_{2a}, q_{2b}, q_A\}$
$q_A \rightarrow$	$\emptyset$	$\emptyset$	$\{q_A\}$	$\{q_0, q_{0a}, q_{0b}, q_{2a}, q_{2b}, q_A\}$

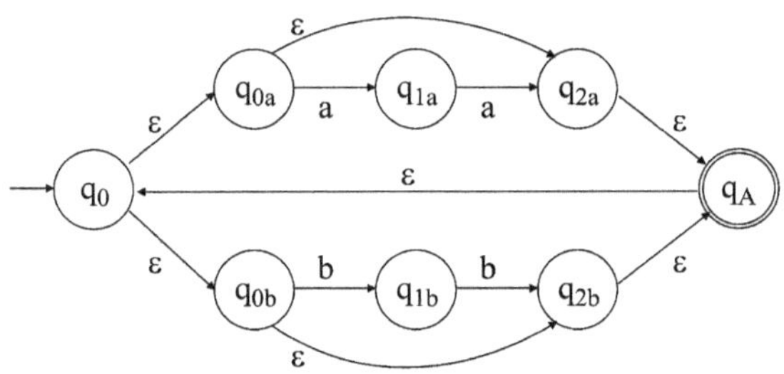

Figure 7.8: The transition diagram of the deterministic finite automaton with ε-transitions given in Example 7.5.

   b. the head of the input tape is placed over the first (leftmost) symbol of the input word,
   c. the control unit is in the initial state q_0,
2. based on a state q of the control unit and on an input symbol $a \in \Sigma$:
   a. $\varepsilon Cl(q)$ is computed,
   b. a state $q' \in \varepsilon Cl(q)$ is nondeterministically picked up,
   c. a set of states $\{p_1, p_2, \ldots, p_k\} = \delta(q', a)$ is computed,
   d. if the set yielded by a transition function is empty then computation is terminated and an input is rejected, otherwise
   e. a state $p' \in \{p_1, p_2, \ldots, p_k\}$ is nondeterministically picked up,
   f. $\varepsilon Cl(p')$ is computed,
   g. a state $p \in \varepsilon Cl(p')$ is nondeterministically picked up,
   h. the input head is shifted right,
   i. the control unit switches to the state p,
3. if the input is nonempty then computation proceeds to the point 2, otherwise computation is terminated and an automaton responds a state of the control unit,
4. it is assumed, like in the case of nondeterministic Turing machines, that a choices of states made in the second point leads to such a computation (in the sense of deterministic computation), which terminates in an accepting state if such a computation exists.

As in previous cases, a fundamental description of a computation of a pushdown automata is strictly consistent with the corresponding description for Turing machines outlined in Definition 5.18 and Remark 5.4.

Like for deterministic and nondeterministic finite automata, computation of a finite automaton with ε-transitions always terminates.

The symbol $\succ$ denotes a transition (of a finite automaton with ε-transitions) in terms of point 2. of Remark 7.10. The symbols $\succ_M$ and $\succ^k$ are used in a usual way, cf. deterministic and nondeterministic finite automata.

Definition 7.13. A binary relation in the space of all possible configurations of the automaton with ε-transition is created by transitions of the automaton (both: transitions related to input symbol reading and ε-transitions). The symbol $\succ^*$ denotes the transitive closure of the transition relation.

Definition 7.14. An interpretation of computation of a finite automaton with ε-transitions $M = (Q, \Sigma, \delta, q_0, F)$ is a tree such that:
- its nodes are labeled by configurations of an automaton,
- its root is labeled by the initial configuration,
- levels created by nodes are distinguished, the root creates the 0th level, children of the root create the 1st level etc.,
- nodes of an odd level are yielded by ε closure applied to nodes of the previous level,
- nodes of an even level are yielded by transition function applied to nodes of the previous level,
- the bottom level has an odd number.

Remark 7.11. Note that, like for nondeterministic finite automata, a computation tree of a finite automaton with ε-transitions is a k-tree, where k is the maximal number of values yielded by the transition function for given arguments (input symbols or the empty word). Moreover, edges of every level of such a tree are labeled with the same input symbol or with the empty word.

Nevertheless, a fundamental description of a computation of a finite automaton with ε-transitions is strictly consistent with the corresponding description for Turing machines outlined in Definition 5.18 and Remark 5.4.

Remark 7.12. Formal definitions of acceptation of input and of a language accepted by a finite automaton with ε-transitions are identical with respective definitions for deterministic and nondeterministic finite automata, that is, Definition 7.5 and Definition 7.6.

Example 7.6. Find a computation tree of a finite automaton with ε-transitions built in Example 7.5 for sample words.

Solution. Computation trees for the empty word ε and for the word $w = aa$ are displayed in Figure 7.9. Both words are accepted because in every tree, a path from the root to an accepting state exists.

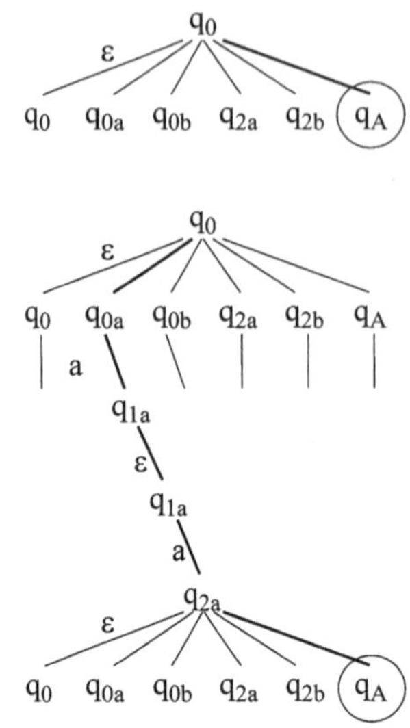

Figure 7.9: Computation trees of the automaton designed in Example 15 for the empty word ε and the word $w = aa$.

Note that transitions of a finite automaton with ε-transitions and a nondeterministic finite automaton are essentially different, cf. point 2 of Remark 7.3 and Remark 7.10. As a result, the *closure of a transition function* of finite automata with ε-transitions differs basically from the closure of a transition function of deterministic and nondeterministic finite automata. The following definition provides details of the closure of transition function for finite automata with ε-transitions.

Definition 7.15. The closure of a transitions function δ of a given finite automaton with ε-transitions $M = (Q, \Sigma, \delta, q_0, F)$ is the function:

$$\hat{\delta} : Q \times \Sigma^* \to 2^Q$$

1. $(\forall q \in Q)\hat{\delta}(q, \varepsilon) = \varepsilon Cl(q)$
2. $(\forall q \in Q)(\forall a \in \Sigma)(\forall w \in \Sigma^*)\hat{\delta}(q, wa) = \varepsilon Cl(\delta(\hat{\delta}(q, w), a))$

where: $\delta(P, a) = \bigcup_{p\in P} \delta(p, a)$ and $\varepsilon Cl(P, a) = \bigcup_{p\in P} \varepsilon Cl(p, a)$ for $P \subset Q$.

As in the case of deterministic and nondeterministic automata, the closure of a transition function of a finite automaton with ε-transitions extends a domain of a transition function from an input alphabet to the set of all words over the input alphabet. The closure of a transition function applied to an initial configuration of a finite automaton immediately yields a result of computation $\hat{\delta}(q_0, w)$ of this automaton for the initial state q_0 and an input word $w = a_1 a_2 \ldots a_n$. However, a restriction of the closure of a transition function $\hat{\delta}$ to the domain $Q \times \Sigma$ is not equal to δ. In fact

$$\hat{\delta}|_{Q\times\Sigma} = \varepsilon Cl(\delta(\varepsilon Cl(q), a))$$

For that reason, a transition function of a finite automaton with ε-transitions and its closure are denoted by different symbols, δ and $\hat{\delta}$, respectively.

Remark 7.13. An input word w is accepted by a finite automaton with ε-transitions $M = (Q, \Sigma, \delta, q_0, F)$ if and only if $\hat{\delta}(q_0, w) \cap F \neq \varnothing$.

A question may be asked whether finite automata with ε-transitions are equivalent to deterministic ones. The answer is positive. Below we present a proof that finite automata with ε-transitions are equivalent to nondeterministic finite automata. Equivalence with deterministic finite automata comes in the form of a direct conclusion of Proposition 7.2.

It is clear that a nondeterministic finite automaton is such a finite automaton with ε-transitions, which does not use ε-transitions. Formally, a nondeterministic finite automaton can be turned to a finite automaton with ε-transitions by extending a domain of its transition function to $Q \times (\Sigma \cup \{\varepsilon\})$ and setting a set of the values of ε-transitions to the empty set.

On the other hand, we will prove that, for a given finite automaton with ε-transitions, there exists an equivalent nondeterministic one, that is accepting the same language. An idea of design of such a nondeterministic automaton is based on analysis of a description of a transition given in point 2 of Remark 7.10. It could be concluded that a transition, as described there, corresponds to a transition of a nondeterministic finite automata. On the other hand, a transition described there is what closure of a transition function for an input symbol yields. Details are given as follows:

Proposition 7.2. *Finite automata with ε-transitions are equivalent to nondeterministic ones.*

Proof. First, the following nondeterministic finite automaton:

$$M = (Q, \Sigma, \delta, q_0, F)$$

is equivalent to an automaton with ε-transitions:

$$M_\varepsilon = (Q, \Sigma, \delta', q_0, F)$$

such that $\delta' : Q \times (\Sigma \cup \{\varepsilon\}) \to 2^Q$, $\delta'(q, a) = \delta(q, a)$, $\delta'(q, \varepsilon) = \varnothing$ for $q \in Q$ and $a \in \Sigma$.

Second, let us assume that now a finite automaton with ε-transitions:

$$M_\varepsilon = (Q, \Sigma, \delta, q_0, F)$$

The equivalent nondeterministic automaton is denoted as follows:

$$M' = (Q, \Sigma, \delta', q_0, F')$$

where:

- $\delta' \equiv \hat{\delta}|_{Q\times\Sigma}$ – the transition function of M' is equal to the closure of the transition function of M restricted to $Q \times \Sigma$,
-

$$F' = \begin{cases} F \cup \{q_0\} & \text{if } \varepsilon Cl(q_0) \cap F \neq \varnothing, \\ F & \text{otherwise.} \end{cases}$$

A formal proof of equivalence of M_ε and M' is based on mathematical induction with regard to the length of input word of both automata.

Let us prove that the closure of transition function of both automata is equal for any nonempty word $w \in \Sigma^+$

$$\delta'(q_0, w) = \hat{\delta}(q_0, w) \tag{$*$}$$

Note that δ' denotes the transition function of M' and its closure. However, the transition function of M_ε and its closure cannot be denoted by the same symbol.

1. for words of length 1 equality $(*)$ is derived directly from definition of both functions,
2. let us check if equality $(*)$ holds for a word $a_1\, a_1\, \dots\, a_n\, a = w\, a$:

$$\delta'(q_0, wa) \underbrace{=}_{\substack{\text{from closure}\\ \text{of transition}\\ \text{function } \delta'}} \bigcup_{p\in\delta'(q_0,w)} \delta'(p,a) \underbrace{=}_{\substack{\text{from}\\ \text{inductive}\\ \text{assumption}}} \bigcup_{p\in\hat{\delta}(q_0,w)} \delta'(p,a)$$

$$\underbrace{=}_{\substack{\text{from}\\ \text{definition}\\ \text{of } \delta'}} \bigcup_{p\in\hat{\delta}(q_0,w)} \hat{\delta}(p,a) \underbrace{=}_{\substack{\text{from closure of}\\ \text{transition}\\ \text{function } \hat{\delta}}} \hat{\delta}(q_0, wa),$$

3. utilizing mathematical induction to 1. and 2. we conclude that the equality $(*)$ holds.

We have just confirmed that both automata M_ε and M' compute the same set of states for given input word $w \in \Sigma^+$. This is only a step to finalize the proof, that is, to show that a given input word $w \in \Sigma^*$ is either accepted by both automata M_ε and M', or is rejected by them. Let us consider the following cases:

- the empty word $w = \varepsilon$ at input. Transition functions yield the following values in this case: $\hat{\delta}(q_0, \varepsilon) = \varepsilon Cl(q_0)$ and $\delta'(q_0, \varepsilon) = \{q_0\}$. Thus, ε is accepted by M_ε if and only if $\hat{\delta}(q_0, \varepsilon) \cap F \neq \varnothing$. On the other hand, $q_0 \in F'$ if and only if $\varepsilon Cl(q_0) \cap F \neq \varnothing$. Since ε is accepted by M' if and only if $q_0 \in F'$, then the empty word ε is either accepted by both automata, or is rejected by them,
- a nonempty word $w \in \Sigma^+$ at input and $\varepsilon Cl(q_0) \cap F = \varnothing$ or $q_0 \in F$. In such the case $F' = F$. Joint acceptation is derived from the equality $(*)$ and definitions of acceptation of an input word by M_ε and M',
- a nonempty word $w \in \Sigma^+$ at input and $\varepsilon Cl(q_0) \cap F \neq \varnothing$ and $q_0 \notin F$. Then for any nonempty word $w \in \Sigma^+$ we get two subcases:

- if $q_0 \notin \delta'(q_0, w)$, then either both $\delta'(q_0, w)$ and $\hat{\delta}(q_0, w)$ include an accepting state, or none does. Again, a joint acceptation is derived from the equality $(*)$ and the definitions of acceptance of an input word by M_ε and M',
- if $q_0 \in \delta'(q_0, w)$ than the word w is accepted by M'. Also $\delta'(q_0, w) = \hat{\delta}(q_0, w)$ due to $(*)$. We have $\varepsilon Cl(\hat{\delta}(q_0, w)) \cap F \neq \varnothing$, because $q_0 \in \hat{\delta}(q_0, w)$. Let us assume that $w = ua$ for $u \in \Sigma^*$ and $a \in \Sigma$. Then, from Definition 7.15 and idempotency of εCl we conclude that some accepting state belongs to $\hat{\delta}(q_0, w)$ due to:

  $$\varepsilon Cl(\hat{\delta}(q_0, w)) = \varepsilon Cl(\varepsilon Cl(\delta(\hat{\delta}(q_0, u), a))) = \varepsilon Cl(\delta(\hat{\delta}(q_0, u), a)) = \hat{\delta}(q_0, w)$$

  and finally, since $\hat{\delta}(q_0, w) = \delta'(q_0, w) \cap F \neq \varnothing$, we derive that the word w is accepted by M_ε,
- no other case can be found.

We have proved that a word $w \in \Sigma^*$ is either jointly accepted by both automata or is jointly rejected by them. This completes the proof. □

Example 7.7. Design a nondeterministic finite automaton equivalent to the automaton shown in Example 7.5. Apply the design method used in the proof of Proposition 7.2.

Solution. The transition table of the nondeterministic finite automaton

$$M = (\{q_0, q_{0a}, q_{1a}, q_{2a}, q_{0b}, q_{1b}, q_{2b}, q_A\}, \Sigma = \{a, b\}, \delta_N, q_0, F = \{q_A\})$$

is shown in Table 7.7. The transition diagram of the automaton is shown in Figure 7.10. There are numerous transitions in the nondeterministic automaton. At a glance, it becomes clear that many states and transitions are duplicates unnecessary for acceptance of the language.

Table 7.7: Transition table of the nondeterministic finite automaton equivalent to the automaton with ε-transitions designed in Example 7.5.

δ_N	a	b
$\rightarrow q_0 \rightarrow$	$\{q_{1a}\}$	$\{q_{1b}\}$
q_{0a}	$\{q_{1a}\}$	$\{q_{1b}\}$
q_{1a}	$\{q_0, q_{0a}, q_{0b}, q_{2a}, q_{2b}, q_A\}$	$\emptyset$
q_{2a}	$\{q_{1a}\}$	$\{q_{1b}\}$
q_{0b}	$\{q_{1a}\}$	$\{q_{1b}\}$
q_{1b}	$\emptyset$	$\{q_0, q_{0a}, q_{0b}, q_{2a}, q_{2b}, q_A\}$
q_{2b}	$\{q_{1a}\}$	$\{q_{1b}\}$
$q_A \rightarrow$	$\{q_{1a}\}$	$\{q_{1b}\}$

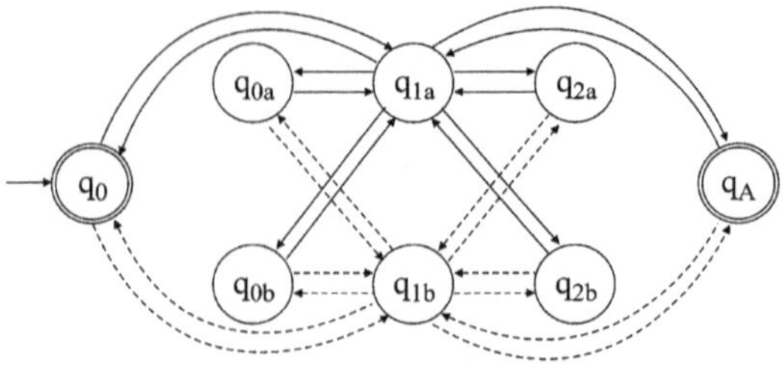

Figure 7.10: Transition diagram of a nondeterministic finite automaton equivalent to the automaton with ε-transitions designed in Example 7.5. Solid edges identify transitions by letter a, dashed edges identify transitions by letter b.

Transition table of the deterministic finite automaton

$$M = (\{q_0, q_{1a}, q_{1b}, q_A, \varnothing\}, \Sigma = \{a, b\}, \delta_D, q_0, F = \{q_A\})$$

is shown Table 7.8. The transition diagram of the automaton is shown in Figure 7.11. The set of useful states is significantly reduced compared to the nondeterministic automaton. Based on Myhill–Nerode Theorem discussed in Chapter 8, we can prove that this is a minimal deterministic automaton accepting this language, minimal in terms of the number of states.

Table 7.8: Transition table of the nondeterministic finite automaton equivalent to the automaton with ε-transitions designed in Example 7.5. Notice that $[q_{0,0a,0b,2a,2b,A}]$ identifies $\{q_0, q_{0a}, q_{0b}, q_{2a}, q_{2b}, q_A\}$.

δ_D	a	b
$\rightarrow [q_0] \rightarrow$	$[q_{1a}]$	$[q_{1b}]$
$[q_{1a}]$	$[q_{0,0a,0b,2a,2b,A}]$	$\emptyset$
$[q_{1b}]$	$\emptyset$	$[q_{0,0a,0b,2a,2b,A}]$
$[q_{0,0a,0b,2a,2b,A}] \rightarrow$	$[q_{1a}]$	$[q_{1b}]$
$\emptyset$	$\emptyset$	$\emptyset$

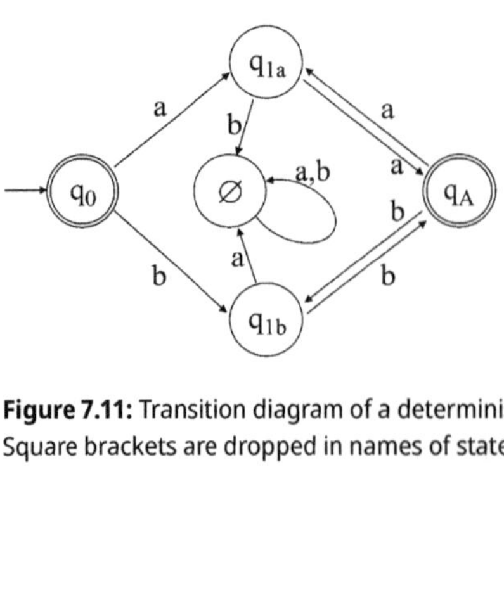

Figure 7.11: Transition diagram of a deterministic finite automaton with transition table given in Table 7.8. Square brackets are dropped in names of states. q_A identifies $\{q_0, q_{0a}, q_{0b}, q_{2a}, q_{2b}, q_A\}$.

7.4 Finite automata as Turing machines

In this section, we justify that finite automata are restricted Turing machines. Furthermore, since finite automata always terminate their computation, they are restricted Turing machines with the stop property. A discussion will be focused on deterministic finite automata. However, since classes of nondeterministic automata and automata with ε-transitions one are equivalent to the class of deterministic ones, the conclusions of this discussion concern all three classes.

A deterministic finite automaton can be simulated by a Turing machine, which shifts the head right and terminates computation as soon as an input word has been read. The machine accepts its input if and only if the automaton accepts it. The details of the design of a Turing machine equivalent to a deterministic finite automaton are given below.

Proposition 7.3. *There exists a Turing machine with the stop property equivalent to a given deterministic finite automaton.*

Proof. Let a deterministic finite automaton is given

$$M = (Q, \Sigma, \delta, q_0, F)$$

A Turing machine in basic model and with halting states equivalent to this deterministic finite automaton is:

$$M_T = (Q_T, \Sigma, \Gamma_T, \delta_T, q_0, B, \{q_A\}, \{q_R\})$$

where:
- $Q_T = Q \cup \{q_A, q_R\}$ – two halting states q_A and q_R are added to the set of states $Q, q_A, q_R \notin Q$,
- $\Gamma_T = \Sigma \cup \{B\}$ – the blank symbol B and the input alphabet create the tape alphabet, $B \notin \Sigma$,
- the transition function δ_T is described with the conditions:

$$\delta_T(q, a) = (\delta(q, a), B, R) \quad \text{for } q \in Q, a \in \Sigma,$$

$$\delta_T(q, B) = \begin{cases} (q_A, B, R) & \text{for } q \in F \\ (q_B, B, R) & \text{for } q \in Q - F \end{cases}$$

The input configuration of the Turing machine M_T is:
- an input word of the deterministic finite automaton M is stored on the tape of M_T,
- the head of the tape is placed over the first cell, that is, on the first input symbol,
- the control unit of M_T is in the initial state q_0. □

Of course, the Turing machine M_T terminates computation as soon as it reaches a halting state. It accepts its input if and only if the deterministic finite automaton M accepts this input.

Finite automata are special cases of pushdown automata:

Proposition 7.4. *There exists a pushdown automaton equivalent to a given deterministic finite automaton.*

Proof. Let a deterministic finite automaton is given as:

$$M = (Q, \Sigma, \delta, q_0, F)$$

A deterministic pushdown automaton equivalent to this deterministic finite automaton is:

$$M_S = (Q, \Sigma, \Gamma, \delta_S, q_0, F)$$

where:
- $\Gamma = \{\triangleright\}$ – there is only one stack symbol, the initial stack symbol,
- the transition function δ_S is described with the condition: $\delta_S(q, a, \triangleright) = \{(\delta(q, a), \triangleright)\}$.

□

7.5 Problems

Problem 7.1. Consider the automaton with the transition function given in the Table 7.9 as a solution of Example 7.4

Table 7.9: Another transition function of the finite automaton designed in Example 7.4.

δ	0	1
$\to q_0$	$\{q_0, q_{10}\}$	$\{q_0, q_{11}\}$
q_{10}	$\{q_{20}\}$	$\emptyset$
q_{20}	$\{q_A\}$	$\emptyset$
q_{11}	$\emptyset$	$\{q_{21}\}$
q_{21}	$\emptyset$	$\{q_A\}$
$q_A \to$	$\emptyset$	$\emptyset$

Problem 7.2. Build a deterministic finite automaton accepting binary numbers divisible by 5.

Hint: Enumerate states of the automaton with remainders from division by 5. To find transitions, consider remainder from division a binary number n by 5 and remainder from the division by 5 a number n with a binary digit attached.

Problem 7.3. Design a finite automaton accepting binary words having no more than four 1s in every sequence of seven consecutive letters.

Problem 7.4. Let $L_1 = L(M_1)$ and $L_2 = L(M_2)$ are languages accepted by deterministic finite automata M_1 and M_2. Build a finite automaton, which accepts:

- union of L_1 and L_2, that is, the language $L = L_1 \cup L_2$,
- concatenation of L_1 and L_2, that is, the language $L = L_1 \circ L_2$,
- intersection of L_1 and L_2, that is, the language $L = L_1 \cap L_2$,
- Kleene closure of L_1, that is, the language $L = (L_1)^*$,
- complement of L_1, that is, the language $L = \Sigma^* - L_1$, where Σ is an alphabet of L_1.

Solution. A design employs a *parallel* computations of both automata M_1 and M_2 for union and intersection, a *sequential* computations of them for concatenation, a *multi-sequential* computations of one automaton for Kleene closure and a normal computation of it for complement.

Let assume that $M_1 = (Q_1, \Sigma, \delta_1, q_0^1, F_1)$ and $M_2 = (Q_2, \Sigma, \delta_2, q_0^2, F_2)$ are deterministic finite automata with disjoint sets of states $Q_1 \cap Q_2 = \varnothing$. The following automata $M = (Q, \Sigma, \delta, q_0, F)$ accept given languages, where:

- for union and intersection:
  - $Q = Q_1 \times Q_2$,
  - $\delta((q', q''), a) = (\delta_1(q', a), \delta_2(q'', a))$ for $a \in \Sigma$, $q' \in Q_1$ and $q'' \in Q_2$, this transition function simulates parallel computation of both automata,
  - $q_0 = (q_0^1, q_0^2)$,
  - for union $F = F_1 \times Q_2 \cup Q_1 \times F_2$, that is, M accepts if and only if at least one automaton accepts,
  - for intersection $F = F_1 \times F_2$, that is, M accept when both automata accept,

  note that the automaton M is a deterministic one,
- for concatenation:
  - $Q = Q_1 \cup Q_2$,
  - $\delta(q, a) = \{\delta_1(q, a)\}$, for $a \in \Sigma$, $q \in Q_1$, M simulates computation of M_1 for a first part of an input,
  - $\delta(q, a) = \{\delta_2(q, a)\}$, for $a \in \Sigma$, $q \in Q_2$, M simulates computation of M_2 for a second part of input,
  - $\delta(q, \varepsilon) = \{q_0^2\}$, for $q \in F_1$, this is a ε-transition, which transfers computation from M_1 to M_2, assuming that M_1 accepts a first part of an input,
  - $q_0 = q_0^1$, M starts computing simultaneously with M_1,
  - $F = F_2$, accept when the second part of an input is accepted (a first part has already been accepted, cf. ε-transitions from F_1),
- $\overline{M} = (Q_1, \Sigma, \delta, q_0, Q_1 - F_1)$ accept complement of $L_1 = L(M_1)$,
- $M' = (Q_1, \Sigma, \delta', q_0, F_1)$ accept $(L_1)^*$, where
  - $\delta'(q, a) = \{\delta(q, a)\}$ for $q \in Q_1$ and $a \in \Sigma$, realizes computation of M_1 for parts of an input word,
  - $\delta'(q_0, \varepsilon) = F_1 - \{q_0\}$, accepts the empty word,
  - $\delta'(q, \varepsilon) = \{q_0\}$ for $q \in F_1 - \{q_0\}$, when the previous part of an input has been accepted, runs computation of M_1 for next part of an input.

Problem 7.5. A language L is accepted by a deterministic finite automaton $M = (Q, \Sigma, \delta, q_0, F)$. Let $L' = \{w = a_1\, a_2\, a_3\, a_4\, \ldots\, a_n \in L : w$ includes all letters of the alphabet $\Sigma\}$.

Hint: Design a deterministic finite automaton M_Σ accepting the language L_Σ of words which include all letters of the alphabet Σ. Then build a finite automaton accepting intersection of both languages L and L_Σ, cf. Problem 7.4.

Problem 7.6. A language L is accepted by a finite automaton. Prove that the language $L^R = \{w : w^R \in L\}$ is accepted by a finite automaton.

Solution. Notice that if $w = a_1\, a_2 \ldots a_k$, then $w^R = a_k \ldots a_2\, a_1$. An idea is to reverse the computation $q_0\, a_1\, q_{i_1}\, a_2\, q_{i_2}\, \cdots\, q_{i_{k-1}}\, a_k\, q_i$ for the word w in a deterministic automaton $M = (Q, \Sigma, \delta, q_0, F)$. Thus, an automaton accepting L^R has reversed transitions, the set of accepting states includes only q_0, an extra state plays role of the initial state with ε-transitions from it to states from F. The formal description of this automaton is $M^R = (Q \cup \{q'\}, \Sigma, \delta^R, q', \{q_0\})$ for $q' \notin Q$ and

- $\delta^R(q, a) = \{p \in Q : \delta(p, a) = q\}$ for $q \in Q$, $a \in \Sigma$,
- $\delta^R(q', \varepsilon) = F$,
- $\delta^R(q', a) = \varnothing$ for $a \in \Sigma$,
- $\delta^R(q, \varepsilon) = \varnothing$ for $q \in Q$.

The reader may formally prove equivalence of both automata.

Problem 7.7. A language L is accepted by a finite automaton. Prove that the language $L' = \{a_1\, a_2\, a_3\, a_4\, \ldots\, a_{2k-1}\, a_{2k} : a_2\, a_1\, a_4\, a_3\, \ldots\, a_{2k}\, a_{2k-1} \in L\}$ is accepted by a finite automaton.

Problem 7.8. A language L is accepted by a finite automaton. Prove that the language $L' = \{w \in L : |w|$ is even$\}$ is accepted by a finite automaton.

Problem 7.9. A language L is accepted by a finite automaton. Prove that the language $L' = \{a_1\, a_1\, a_2\, a_2\, a_3\, a_3\, \ldots\, a_{k-1}\, a_{k-1}\, a_k\, a_k : a_1\, a_2\, a\, a_3\, \ldots\, a_{k-1}\, a_k \in L\}$ is accepted by a finite automaton.

Problem 7.10. A language L is accepted by a finite automaton. Prove that the language of words of L with even letters removed $L' = \{a_1\, a_3\, \ldots\, a_{2k-1} : a_1\, a_2\, a_3\, a_4\, \ldots\, a_{2k-1}\, a_{2k} \in L \vee a_1\, a_2\, a_3\, a_4\, \ldots\, a_{2k-1} \in L\}$ is accepted by a finite automaton.

Part III: **Revisited: languages, grammars, automata**

8 Grammars versus automata

In Chapter 2 we introduced regular expressions and regular grammars, which are tools used to generate regular languages. Then, in Chapter 7 we defined and discussed finite automata.

In this Chapter, we integrate notions of grammars and automata. Specifically, we show that regular expressions, regular grammars and finite automata are equivalent. They generate (expressions and grammars) and accept (automata) the same family of languages, i. e., regular languages. Also, we prove the pumping lemma for regular languages. Let us recall that this lemma was formulated in Chapter 2 and then used as a tool to process languages. Then we formulate and prove the most profound issue concerning regular languages, i. e., the Myhill–Nerode Theorem. As in the case of the pumping lemma, we formulated a part of this theorem in Chapter 2, called the Myhill Nerode Lemma, and used it as a tool for processing languages. Then we discuss the equivalence of context-free grammars and pushdown automata. Without formal proofs, we also briefly discuss the equivalence of unrestricted grammars and Turing machines and context-sensitive grammars and linear bounded automata.

8.1 Regular expressions, regular grammars and finite automata

8.1.1 Regular expressions versus finite automata

Below we prove that regular expressions are equivalent to finite automata. First, we design automata equivalent to regular expressions. The proof is based on the inductive definition of regular expressions, cf. Definition 2.1, and applies mathematical induction. A designed finite automaton equivalent to a given regular expression is an automaton with ε-transitions. Then, a regular expression equivalent to a deterministic finite automaton is designed. Furthermore, the equivalence of finite automata and regular expression is drawn based on the above two constructs and on the equivalence of all three classes of finite automata as shown in Figure 8.1.

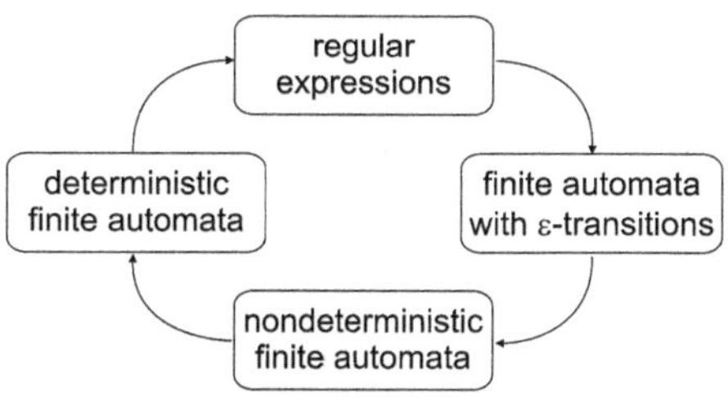

Figure 8.1: Equivalence of finite automata and regular expressions, a dependency diagram.

https://doi.org/10.1515/9783112223123-011

Proposition 8.1. *Languages generated by regular expressions are accepted by finite automata.*

Proof. We use mathematical induction to prove this proposition. A formal proof of equivalence is done concerning the length of a regular expression, i. e., the number of symbols in it.

1. There are 3 families of regular expressions of length 1:
   a. the empty regular expression Θ generates the empty language $\emptyset$. An equivalent finite automaton is shown in Figure 8.2(a)
   b. the empty regular expression ε generates the language with the empty word $\{\varepsilon\}$. An equivalent finite automaton is shown in Figure 8.2(b)
   c. a family of regular expressions $\boldsymbol{a}$, for each symbol of an input alphabet $a \in \Sigma$, generate languages with a one letter word $\{a\}$. An equivalent finite automaton for a given regular expression $\boldsymbol{a}$ is shown in Figure 8.2(c)

   Note that automata shown in Figure 8.2 are nondeterministic ones.
2. Assume that two regular expressions s and t are given and that these expressions generate languages S and T, respectively. Assume that both expressions are given in the same alphabet Σ. Otherwise, take the union of alphabets of both expressions as a joint alphabet.
   Based on inductive assumption, take finite automata M_S and M_T shown in Figure 8.3. Assume that these automata are equivalent to regular expressions s and t, respectively. Now, we consider sum $(s + t)$, concatenation $(s \circ t)$ and Kleene closure (s^*) of regular expressions. Languages generated by these expressions are $S \cup T$, $S \circ T$ and S^*. Finite automata (with ε-transitions), which accept union and concatenation of languages S and T, and a finite automaton accepting Klenee closure of the language S are designed as shown in Figures 8.4, 8.5 and 8.6, respectively.
   Notice, that those automata have ε-transitions. Namely, an arrow from the set of states F_S to the state q_A in Figure 8.4 represents ε-transitions from each state of the set F_S to the state q_A. Alike, an arrow from the set of states F_T and the state q_A or to the state q_0^T in Figures 8.4, 8.5 and 8.6.
   A formal proof of equivalence for the union of languages is given in Remark 8.1. Proofs for concatenation and Kleene closure are left to the reader.
3. Employing mathematical induction to 1. and 2. we conclude that there exists a finite automaton equivalent to any regular expression. □

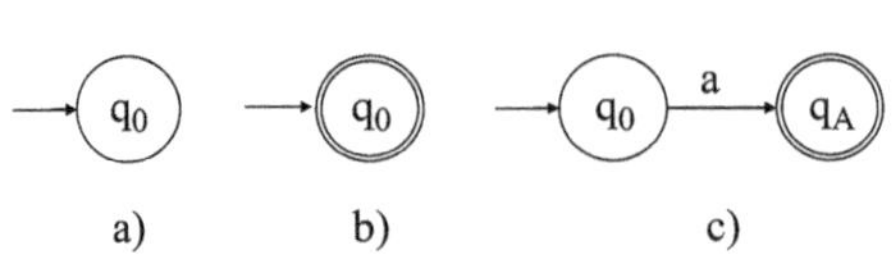

Figure 8.2: Finite automata equivalent to basic regular expressions.

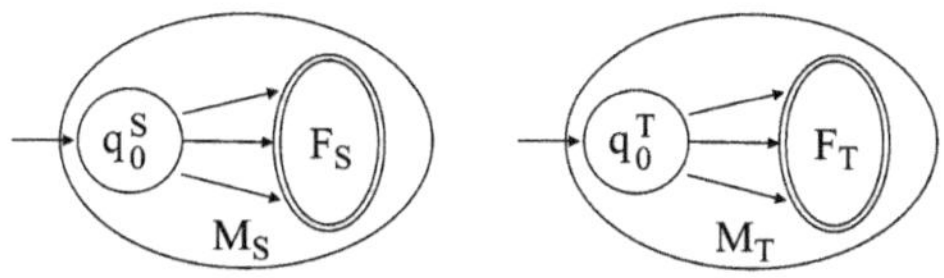

Figure 8.3: Finite automata M_S and M_T equivalent to given regular expressions s and t.

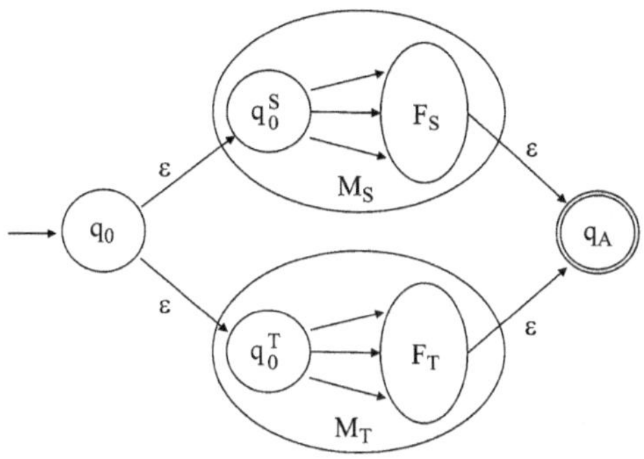

Figure 8.4: A finite automaton equivalent to sum of regular expressions s and t.

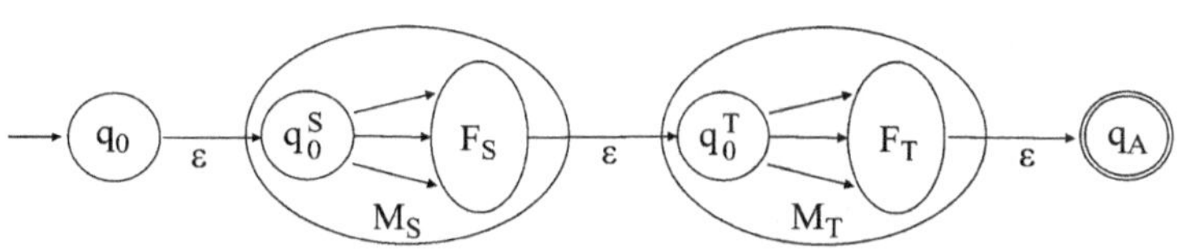

Figure 8.5: A finite automaton equivalent to concatenation of regular expressions s and t.

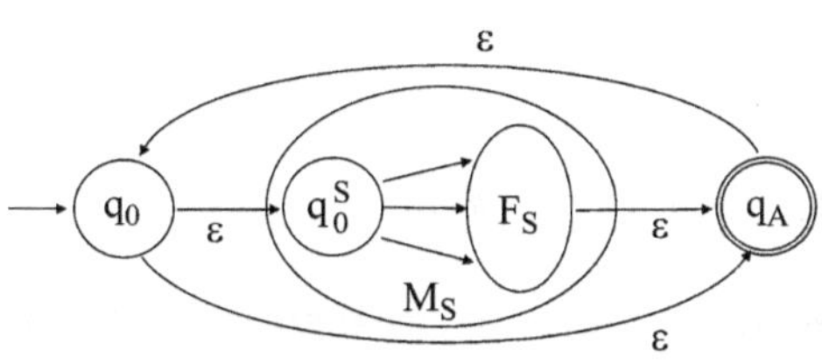

Figure 8.6: A finite automaton equivalent to Kleene closure of a regular expression s.

Remark 8.1. Let $M_S = (Q_S, \Sigma, \delta_S, q_0^S, F_S)$ and $M_T = (Q_T, \Sigma, \delta_T, q_0^T, F_T)$. The finite automaton shown in Figure 8.4 is $M = (Q, \Sigma, \delta, q_0, F)$, where:

- $Q = Q_S \cup Q_T \cup \{q_0, q_A\}$, assuming that sets of states are pairwise disjoint, i. e., $Q_S \cap Q_T = \emptyset$, $Q_S \cap \{q_0, q_A\} = \emptyset$ and $Q_T \cap \{q_0, q_A\} = \emptyset$,
- $F = \{q_A\}$,
- the transitions function δ is based on transition functions of both automata Q_S and Q_T:
  - $\delta(q, a) = \delta_S(q, a)$ for $q \in Q_S$ and $a \in \Sigma$,

- $\delta(q, a) = \delta_T(q, a)$ for $q \in Q_T$ and $a \in \Sigma$,
- $\delta(q, \varepsilon) = \delta_S(q, \varepsilon)$ for $q \in Q_S \setminus F_S$,
- $\delta(q, \varepsilon) = \delta_T(q, \varepsilon)$ for $q \in Q_T \setminus F_T$,
- $\delta(q, \varepsilon) = \delta_S(q, \varepsilon) \cup \{q_A\}$ for $q \in F_S$,
- $\delta(q, \varepsilon) = \delta_T(q, \varepsilon) \cup \{q_A\}$ for $q \in F_T$,
- $\delta(q_0, \varepsilon) = \{q_0^S, q_0^T\}$,
- $\delta(q, X) = \emptyset$ for such $(q, X) \in Q \times (\Sigma \cup \{\varepsilon\})$, which are not considered above.

Proposition 8.2. *Languages accepted by finite automata are generated by regular expressions.*

Proof. Let us assume that a deterministic finite automaton

$$M = (\{q_1, q_2, \ldots, q_n\}, \Sigma, \delta, q_1, F)$$

accepts the language $L = L(M)$. A method of designing a regular expression equivalent to the automaton M relies on the construction of families of languages, which are regular and which are easy to get regular expressions generating them.

The families $R_{i,j}^k$ of languages are designed, where $R_{i,j}^k$, for given natural numbers $i, j \geq 1$ and $k \geq 0$, denotes a set of such words $w \in \Sigma^*$, that a computation for a word w, starting in the state q_i, ends in the state $q_j = \delta(q_i, w)$ and does not visit states with indexes greater than k. Note that indexes i and j may be greater than k. Let us recall that a computation of a deterministic finite automaton is a sequence of alternating states and input symbols. In our case, the computations for the word w begins with the state q_i and ends with the state q_j.

Formal description of the family $R_{i,j}^k$ of languages is as follows:

$$R_{i,j}^0 = \begin{cases} \{a \in \Sigma : \delta(q_i, a) = q_j\} & \text{for } i \neq j \\ \{a \in \Sigma : \delta(q_i, a) = q_j\} \cup \{\varepsilon\} & \text{for } i = j \end{cases}$$

$$R_{i,j}^k = R_{i,k}^{k-1} \circ \left(\left(R_{k,k}^{k-1}\right)^*\right) \circ R_{k,j}^{k-1} \cup R_{i,j}^{k-1} \quad \text{for } k > 0$$

Languages of the family $R_{i,j}^0$ include one-letter words, for which there is a transition from the state q_i to the state q_j. The empty word is included in languages $R_{i,i}^0$ according to the rule that the empty word always allows for a transition to the same state. Note that any word longer than one cannot be included in a language of this family because the computation for such the word includes a middle state with the index greater than 0.

Languages of a family $R_{i,j}^k$, for a given $i, j, k > 0$, are assembled according to the following rules:

- a computation for a word w, which does not visit states with indexes greater than k, but visits the state q_k, may be decomposed to computations which do not visit q_k, such that:
  - a computation that begins in q_i and ends in q_k, this computation forms the language $(R_{i,k}^{k-1})$,

  - multiple computations that begin and end in q_k, they form the language $(R^{k-1}_{k,k})^*$,
  - a computation that begins in q_k and ends in q_j, it forms the language $R^{k-1}_{k,j}$.

  Concatenation of languages $R^{k-1}_{i,k}$, $(R^{k-1}_{k,k})^*$ and $R^{k-1}_{k,j}$ form the first part of the union in the second formula,
- a computation for a word w, which does not visit states with index greater than k (except the beginning state q_i and the ending state q_j, which may be equal or even greater than k), may not visit the state q_k. In that case, $w \in R^{k-1}_{i,j}$, what creates the second part of the union in the second formula.

The language $L(M)$ is a set of such words, for which computation begins in q_1, ends in $q \in F$ and may visit any state of M, i. e.

$$L(M) = \bigcup_{\{j:q_j \in F\}} R^n_{1,j}$$

Now, existence of regular expressions generating languages of families $R^k_{i,j}$ can be proved employing mathematical induction with regard to k:
- languages of the family $R^0_{i,j}$ are generated by the following regular expressions and this is a direct conclusion from the definition of regular expressions:
  - for $i \neq j$,
    - if $R^0_{i,j} = \{a_{i_1}, \ldots, a_{i_p}\}$ for $a_{i_1}, \ldots, a_{i_p} \in \Sigma$, then $r^0_{i,j} = a_{i_1} + \cdots + a_{i_p}$,
    - if $R^0_{i,j} = \emptyset$, then $r^0_{i,j} = \Theta$,
  - for $i = j$,
    - if $R^0_{i,j} = \{\varepsilon, a_{i_1}, \ldots, a_{i_p}\}$, $a_{i_1}, \ldots, a_{i_p} \in \Sigma$, then $r^0_{i,j} = \varepsilon + a_{i_1} + \cdots + a_{i_p}$,
    - if $R^0_{i,j} = \{\varepsilon\}$, then $r^0_{i,j} = \varepsilon$,
- based on inductive hypothesis assume that languages of a family $R^{k-1}_{i,j}$ are generated by regular expressions $r^{k-1}_{i,j}$. Notice that the formula for $R^{k-1}_{i,j}$ is an assembly of languages of a family $R^{k-1}_{i,j}$ using union, concatenation and Kleene closure. The assembly operators correspond to operations on regular expressions: sum, concatenation and Kleene closure. These notes lead to the following regular expression generating the $R^k_{i,j}$ language for a given indexes $i, j, k > 0$:

$$r^k_{i,j} = r^{k-1}_{i,k} \circ \left(r^{k-1}_{k,k}\right)^* \circ r^{k-1}_{k,j} + r^{k-1}_{i,j}$$

In conclusion, the following regular expression is equivalent to the automaton M, assuming that $F = \{q_{i_1}, q_{i_2}, \ldots, q_{i_p}\}$:

$$r^n_{1,i_1} + r^n_{1,i_2} + \cdots + r^n_{1,i_p} \qquad \square$$

Example 8.1. Design a deterministic finite automaton accepting the language of non-empty binary words with numbers of 0's and 1's not equal modulo 3. The language is described as $L = \{w \in \{0,1\}^+ : \#_0 w - \#_1 w \neq 0 \bmod 3\}$. Then, find a regular expression equivalent to the given automaton.

Solution. A transition diagram of a deterministic automaton accepting the language L is given in Figure 8.7. Notice that a state q_i informs that the part u of an input word already read satisfy the condition $\#_0 u - \#_0 u = i + 1$.

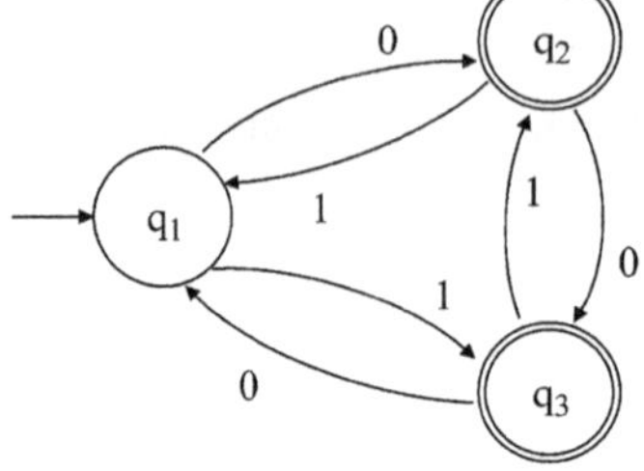

Figure 8.7: A deterministic finite automaton accepting the language given in Example 8.1.

Helper regular expressions necessary for finding a final one equivalent to the automaton are drawn in Table 8.1. Regular expressions with the upper parameter equal to 3 are outlined outside the table due to their size. Note: regular expressions given in Table 8.1 and below are optimized in length with qualities discussed in Section 2.1.1.

$$\begin{aligned} r^3_{1,2} &= r^2_{1,3}(r^2_{3,3})^* r^2_{3,2} + r^2_{1,2} \\ &= (0(10)^*(0+11)+1)(01+(00+1)(10)*(0+11))^*((00+1)(10)^*) + 0(10)^* \\ r^3_{1,2} &= r^2_{1,3}(r^2_{3,3})^* r^2_{3,2} + r^2_{1,2} \\ &= (0(10)^*(0+11)+1)(01+(00+1)(10)*(0+11))^*((00+1)(10)^*) + 0(10)^* \end{aligned}$$

The regular expression equivalent to the finite automaton M is rather complex:

$$\begin{aligned} r_M &= r^3_{1,2} + (r^3_{1,3}) \\ &= (0(10)^*(0+11)+1)(01+(00+1)(10)*(0+11))^*((00+1)(10)^*(\varepsilon+1)+0) \\ &\quad + 0(10)^*(\varepsilon+0+11)+1 \end{aligned}$$

Table 8.1: A helper table for the elaboration of a regular expression equivalent to a finite automaton built in Example 8.1.

	$k=0$	$k=1$	$k=2$	$k=3$
$r^k_{1,1}$	ε	ε	$\varepsilon + 0(10)^*1$	
$r^k_{1,2}$	0	0	$0(10)^*$	$r^2_{1,3}(r^2_{3,3})^* r^2_{3,2} + r^2_{1,2}$
$r^k_{1,3}$	1	1	$0(10)^*(0+11)+1$	$r^2_{1,3}(r^2_{3,3})^* r^2_{3,3} + r^2_{1,3}$
$r^k_{2,1}$	1	1	$(10)^*1$	
$r^k_{2,2}$	ε	$\varepsilon + 10$	$(10)^*$	
$r^k_{2,3}$	0	$0+11$	$(10)^*(0+11)$	
$r^k_{3,1}$	0	0	$(00+1)(10)^*1+0$	
$r^k_{3,2}$	1	$00+1$	$(00+1)(10)^*$	
$r^k_{3,3}$	ε	$\varepsilon + 01$	$\varepsilon + 01 + (00+1)(10)^*(0+11)$	

8.1.2 Regular grammars versus finite automata

Now we prove that regular grammars are equivalent to finite automata. First, we design automata equivalent to right-linear grammars. Then, for a given deterministic finite automaton, an equivalent right-linear grammar is designed. Afterward, it is shown that right-linear grammars are equivalent to left-linear grammars. As a result, we come to a graph of equivalence of regular expressions, regular grammars and finite automata. The graph is displayed in Figure 8.8.

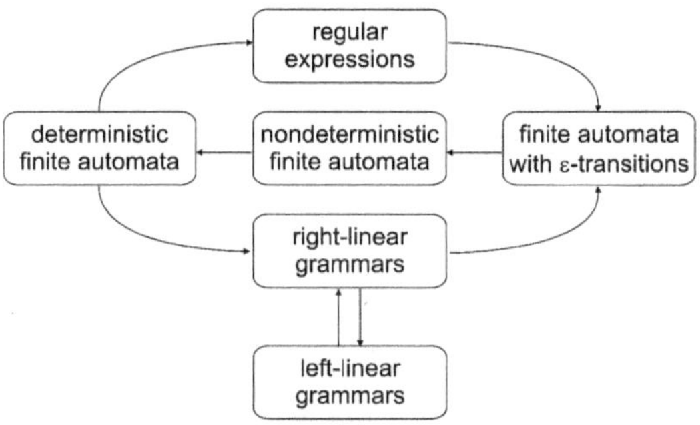

Figure 8.8: Equivalence of finite automata, regular expressions and regular grammars: a dependency diagram.

Theorem 8.1. *Languages accepted by finite automata are generated by right linear grammars.*

Proof. We will prove that for any deterministic automaton M, there is a right linear grammar G_M generating the language $L(M)$, that is, the language accepted by the automaton M. Construction of such a grammar relies on transforming computation of the automaton M into a derivation in designed grammar G_M.

Let us assume that we have the following deterministic finite automaton.

$$M = (Q, \Sigma, \delta, q_0, F)$$

A right-linear grammar G_M, which generates the language $L = L(M)$ accepted by the automaton M, is as follows

$$G_M = (V, T, P, S)$$

where:
- $V = Q \cup \{S\}, Q \cap \{S\} = \emptyset$,
- $T = \Sigma$,
- P – the set of productions includes the following productions and only such productions:
  - $S \to q_0$,
  - $S \to \varepsilon$ if and only if $q_0 \in F$,

- $p \to aq$ if and only if $\delta(p, a) = q$, for all $p, q \in Q$ and $a \in \Sigma$,
- $p \to a$ if and only if $\delta(p, a) \in F$, for all $p \in Q$ and $a \in \Sigma$,

Now, let us prove that a word w is accepted by the automaton M ($w \in L(M)$) if and only if it is generated in the grammar G_M ($w \in L(G_M)$). To prove this, notice that the computation of the automaton M for given the word is identical to this word's derivation in the grammar G_M.

First, we confirm that if a word $w \in L(M)$, then $w \in L(G_M)$. If $\varepsilon \in L(M)$, i. e., $q_0 \in F$, then its derivation is immediate: $S \to \varepsilon$. Now, let us assume that $W \neq \varepsilon$, that is, $w = a_1 a_2 \dots a_n \in L(M) \subset \Sigma^*$ and $n > 0$. Then

$$q_0\ a_1\ q_{i_1}\ a_2\ q_{i_2}\ \cdots\ q_{i_{n-1}}\ a_n\ q_{i_n}$$

is the computation of M for the word w. Since $w \in L(M)$, then $q_{i_n} \in F$. The corresponding derivation in G_M is as follows:

$$S \to q_0 \to a_1\ q_{i_1} \to a_1\ a_2\ q_{i_2} \to \cdots \to a_1\ a_2 \dots a_{n-1}\ q_{i_{n-1}} \to a_1\ a_2 \dots a_{n-1}\ a_n$$

Notice that the production $q_{i_{n-1}} \to a_n$ is included in this grammar because $\delta(q_{i_{n-1}}, a_n) \in F$ and this production is used as the last one in the derivation.

Now, we verify that if $w \in L(G_M)$, then $w \in L(M)$. Let us assume that we have a word $w = a_1 a_2 \dots a_n \in L(G_M)$, then for $w \neq \varepsilon$ there exists a derivation in G_M and it gets the form:

$$S \to q_0 \to a_1\ q_{i_1} \to a_1\ a_2\ q_{i_2} \to \cdots \to a_1\ a_2 \dots a_{n-1}\ q_{i_{n-1}} \to a_1\ a_2 \dots a_{n-1}\ a_n$$

The corresponding computation of M for any $w \neq \varepsilon$ is as follows:

$$q_0\ a_1\ q_{i_1}\ a_2\ q_{i_2}\ \cdots\ q_{i_{n-1}}\ a_n\ q_{i_n}$$

where the production $q_{i_{n-1}} \to a_n q_{i_n}$ used in computation corresponds to the production $q_{i_{n-1}} \to a_n$ employed in the derivation, for some $q_{i_n} \in F$.

Subsequently, for $w = \varepsilon \in L(M)$ there is the production, which constitutes the derivation $S \to \varepsilon$. The corresponding computation is q_0. □

Example 8.2. Design a right-linear grammar equivalent to the finite automaton shown in Figure 8.9.

Solution. The transition graph of the automaton is given in Figure 8.9.

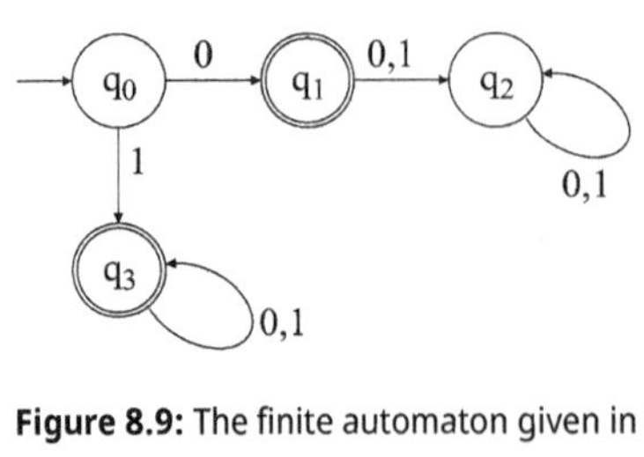

Figure 8.9: The finite automaton given in Example 8.2.

The right-linear grammar designed according to Theorem 8.1 is as follows:

$$G = (\{S, q_0, q_1, q_2, q_3\}, \{0,1\}, P, S)$$

with productions:

$$P : \begin{cases} S \to q_0 \\ q_0 \to 0q_1 \mid 1q_3 \mid 0 \mid 1 \\ q_1 \to 0q_2 \mid 1q_2 \\ q_2 \to 0q_2 \mid 1q_2 \\ q_3 \to 0q_3 \mid 1q_3 \mid 0 \mid 1 \end{cases}$$

This right-linear grammar is a context-free grammar. So, it may be simplified by removing useless symbols. Finally, we get the grammar:

$$G = (\{S, q_3\}, \{0,1\}, P', S)$$

$$P' : \begin{cases} S \to 1q_3 \mid 0 \mid 1 \\ q_3 \to 0q_3 \mid 1q_3 \mid 0 \mid 1 \end{cases}$$

Theorem 8.2. *Languages generated by right-linear grammars are accepted by finite automata.*

Proof. Let

$$G = (V, T, P, S)$$

is a right-linear grammar. A finite automaton with ε-transitions accepts the language $L = L(G)$ generated by the grammar G:

$$M = (Q, \Sigma, \delta, q_0, F)$$

where:
- $Q = \{[\alpha] \in (V \cup T)^* : (\exists \beta \in (V \cup T)^*) A \to \beta\alpha \in P\}$, i. e., states are labeled with all possible suffixes of right hand sides of productions,
- $\Sigma = T$,
- $q_0 = [S]$,
- δ – a transition function is assembled with the rules:
  - if $A \in V$ then $\delta([A], \varepsilon) = \{[\alpha] : A \to \alpha \in P\}$,
  - if $a \in T \wedge \alpha \in (T^* \cup T^*V)$ then $\delta([a\alpha], a) = \{[\alpha]\}$,
- $F = \{[\varepsilon]\}$.

Justification of correctness of the automaton construction is based on observation of a derivation in a right-linear grammar.

Any intermediate derivation word is of a form xA, where $x \in T^*, A \in V$. A right-hand side of a production $A \to yB$ employed in a derivation replaces a nonterminal symbol

A and creates a next intermediate derivation word xyB. An automaton reads terminal symbols inserted by a production and then switches to a state relevant to an inserted nonterminal symbol. This action is repeated for all productions of this form employed in a derivation. A derivation is terminated with a production $A \to z$, where $z \in T^*$ (without a nonterminal symbol on its right-hand side). In this case, an automaton reads terminal symbols and then goes to an accepting state marked with the empty word ε.

Computation of such an automaton, for a given the word, follows a derivation of this word. States visited along a computation correspond to unread part of an input word. This unread part is represented by beginning terminal symbols and by one nonterminal symbol if the derivation is not terminated with a production to a string of terminals employed. This nonterminal symbol produces a remaining part of an input word. The last production of a derivation does not include a nonterminal, which allows an automaton to read terminal symbols and to go to the accepting state.

The above notes can be turned to formal inductive proof with regard to the length of derivation. □

Example 8.3. Design a finite automaton equivalent to the right-linear grammar

$$G = (\{S, A\}, \{0, 1\}, \{S \to 00A,\ A \to 11A \mid 11S \mid 1 \mid \varepsilon\},\ S)$$

Solution. The automaton is designed according to the method used in the proof of Theorem 8.3. A transition diagram of the automaton is shown in Figure 8.10.

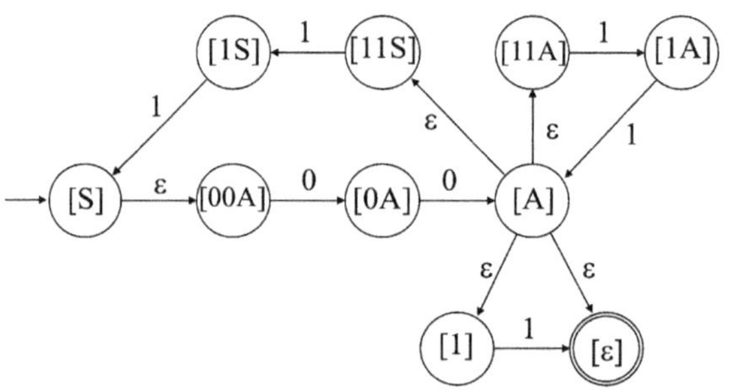

Figure 8.10: A finite automaton equivalent the right-linear grammar given in Example 8.3.

Theorem 8.3. *Right-linear grammars are equivalent to left-linear grammars.*

Proof. In Problem 7.6 it is proved that a language L^R is regular if L is regular, where L^R is the language of reversed words of L. Observe that a right-linear grammar G^R is turned to a left-linear one G_L if right-hand sides of G_R productions are reversed. Moreover, a language $L(G_L)$ generated by G_L is an inverse of $L(G_R)$, i. e., $(L(G_L))^R = L(G_R)$, assuming that G_L is created by reversing productions of G_L. A formal proof of this equality relies on a simple application of mathematical induction with regard to the length of the derivation of words, i. e., with regard to the number of productions employed in a derivation. □

8.1.3 The pumping lemma

The pumping lemma was formulated in Chapter 2, but not proved there. Here it is recalled and proved.

Theorem 8.4 (The pumping lemma for regular languages).
If *a language L is regular*
then *there exists a constant n_L such that for any word $z \in L$ the following condition holds:*

$$(|z| \geq n_L) \Rightarrow \left[\left(\bigvee_{u,v,w} z = uvw \wedge |uv| \leq n_L \wedge |v| \geq 1 \right) \bigwedge_{i=0,1,2,...} z_i = uv^i w \in L \right]$$

Proof. If a language L is a regular one, then there exists a deterministic finite automaton $M = (Q, \Sigma, \delta, q_0, F)$, which accepts L. Let denote the number of states of this automaton $|Q| = n$. If all words of the language L are shorter than n, then for the constant $n_L = n$, the implication holds because its antecedent $|z| \geq n_L$ is never true.

Let $z \in L$, $z = a_1 a_2 \ldots a_m$, where $m \geq n$. A computation of M for z is

$$q_0 a_1 q_{i_1} a_2 q_{i_2} \ldots a_{m-1} q_{m_{m-1}} a_m q_{i_m}$$

and $q_{i_m} \in F$.

There are $m+1 \geq n+1$ states in this computation. This means that at least one state is repeated because there are only n states in M. Let us take the leftmost pair of repeating states. They, of course, appear in a beginning part of the computation, including no more than n letters and no more than $n + 1$ states. In the following computation, leftmost repeating states are underscored

$$q_0 a_1 q_{i_1} a_2 \ldots a_j \underline{q_{i_j}} a_{j+1} q_{i_{j+1}} \ldots a_{j+p} \underline{q_{i_{j+p}}} a_{j+p+1} q_{i_{j+p+1}} \ldots a_{m-1} q_{i_{m-1}} a_m q_{i_m}$$

where $p \geq 1$.

For that reason, the part of computation between these underlined states includes at least one letter. If the first underlined state together with the part between underlined (repeated) states, i. e., $\underline{q_{i_j}} a_{j+1} q_{i_{j+1}} \ldots a_{j+p}$, is removed. The remaining sequence of states and letters is still a computation, which ends in an accepting state:

$$q_0 a_1 q_{i_1} a_2 \ldots a_j \underline{q_{i_{j+p}}} a_{j+p+1} q_{i_{j+p+1}} \ldots a_{m-1} q_{i_{m-1}} a_m q_{i_m}$$

On the other hand, the sequence $\underline{q_{i_j}} a_{j+1} q_{i_{j+1}} \ldots a_{j+p}$ may be inserted just before the first repeated state and an obtained sequence is still a computation, which ends in an accepting state:

$$q_0 a_1 q_{i_1} \ldots a_j \underline{q_{i_j}} a_{j+1} \ldots a_{j+p} \underline{q_{i_j}} a_{j+1} \ldots a_{j+p} \underline{q_{i_{j+p}}} a_{j+p+1} \ldots a_{m-1} q_{i_{m-1}} a_m q_{i_m}$$

Insertion of this sequence may be repeated, developing a computation, which ends in an accepting state.

As a result, the computations for following words are created:

- $z_0 = a_1 a_2 \dots a_j a_{j+p+1} \dots a_{m-1} a_m$,
- $z_1 = a_1 a_2 \dots a_j \underline{a_{j+1} \dots a_{j+p}} a_{j+p+1} \dots a_{m-1} a_m$,
- $z_2 = a_1 a_2 \dots a_j \underline{a_{j+1} \dots a_{j+p} a_{j+1} \dots a_{j+p}} a_{j+p+1} \dots a_{m-1} a_m$,
- $z_3 = a_1 a_2 \dots a_j \underline{a_{j+1} \dots a_{j+p} a_{j+1} \dots a_{j+p} a_{j+1} \dots a_{j+p}} a_{j+p+1} \dots a_{m-1} a_m$,
- etc.

and they end in an accepting state. Notice that a part of a word, which is repeated, has at least one letter. Moreover, both the beginning part and a repeated part are not longer than $n_L = n$. Therefore, we have a sequence of words as required in the consequence of the implication. This proves the lemma. □

As a consequence of the pumping lemma, it may be concluded that computations of a finite automaton are determined by a finite set of words shorter than a constant n_L. A computation for a word, which is not shorter than n_L, can be shortened by removing its middle part(s), as in the pumping lemma. This implies that a set of accepting states of a deterministic finite automaton can be effectively calculated by investigating a finite set of such words, which are not longer than the constant n_L. Computations for longer words cannot bring a new accepting state. This conclusion can be formally expressed as follows.

Remark 8.2. For any word $z \in L$, $|z| \geq n_L$, there exists a word $w \in L$, $|w| < n_L$ such that the computation for the word z is $\alpha_z = \alpha_1 \alpha_2 \alpha_3$ and the computation for the word w is $\alpha_w = \alpha_1 \alpha_3$. Note that α_2 may include many different repeating parts.

8.1.4 The Myhill–Nerode Theorem

In this section the Myhill–Nerode theorem is formulated and proved. The Myhill–Nerode lemma, which was used in Chapter 2, is a direct consequence of the Myhill–Nerode theorem.

Theorem 8.5 (The Myhill–Nerode Theorem). *The following conditions are equivalent:*

1. *a language L is accepted by a deterministic finite automaton* $M = (Q, \Sigma, \delta, q_0, F)$,
2. *a language L is a union of some classes of a right invariant equivalence relation with finite index,*
3. *the relation* R_L *induced by a language L has finite index.*

Proof. The following implications between the above conditions will be shown: $1 \Rightarrow 2 \Rightarrow 3 \Rightarrow 1$.

$1 \Rightarrow 2$

Assume that a deterministic finite automaton $M = (Q, \Sigma, \delta, q_0, F)$ is given. Let us define a relation $\rho_M \subset \Sigma^* \times \Sigma^*$ such that for any $x, y \in \Sigma^*$, $x\rho_M y \Leftrightarrow \delta(q_0, x) = \delta(q_0, y)$. The relation is:

- a right invariant relation because:
  $(\forall x, y, z \in \Sigma^*)\delta(q_0, x) = \delta(q_0, y) \Rightarrow \delta(q_0, xz) = \delta(q_0, yz)$,
- an equivalence relation since the equality relation is an equivalence relation. It is obvious that ρ_M is
  - reflexive, i. e., $(\forall x \in \Sigma^*)\, \delta(q_0, x) = \delta(q_0, x)$,
  - symmetric, i. e., $(\forall x, y \in \Sigma^*)\, \delta(q_0, x) = \delta(q_0, y) \Rightarrow \delta(q_0, y) = \delta(q_0, x)$,
  - transitive, i.e., $(\forall x, y, z \in \Sigma^*)\, \delta(q_0, x) = \delta(q_0, y) \wedge \delta(q_0, y) = \delta(q_0, z) \Rightarrow \delta(q_0, x) = \delta(q_0, z)$.

It is evident that all words for which computation ends in the same state create an equivalence class. In conclusion, we wind up that the number of equivalence classes is not greater than the number of states $|Q|$ of the automaton M. It is also evident that the language L is a union of those equivalent classes, which correspond to accepting states.

$$2 \Rightarrow 3$$

Let us assume that a relation ρ as described in condition 2 is given. For any $x, y \in \Sigma^*$, if $x\rho y$, then either $x, y \in L$, or $x, y \notin L$ (because L is a union of some equivalence classes of ρ). Moreover, for any $z \in \Sigma^*$, if $x\rho y$, then $xz\rho yz$, i. e., either $xz, yz \in L$, or $xz, yz \notin L$, (because is a right invariant relation). For that reason $(\forall x, y \in \Sigma^*)x\rho y \Rightarrow xR_L y$. In conclusion, every equivalence class of the relation ρ is included in some equivalence class of the relation R_L induced by the language L. We get that R_L has no more equivalence classes than ρ_M has, i. e., the number of equivalence classes of R_L is finite.

$$3 \Rightarrow 1$$

Assume that the relation R_L induced by the language L has a finite number of equivalence classes. The following deterministic finite automaton accepts the language L:

$$M = (Q, \Sigma, \delta, q_0, F)$$

where:

- $Q = \{q_{[w]} : w \in \Sigma^*\}$ – states correspond to equivalence classes of R_L,
- Σ – an alphabet of the language L,
- $q_0 = q_{[\varepsilon]}$ – the state corresponding to the equivalence class $[\varepsilon]$, which includes the empty word ε, is the initial state,
- $F = \{q_{[w]} : w \in L\}$ – accepting states correspond to equivalence classes, which are included in the language L,
- δ – a transition function is defined by the formula $\delta(q_{[w]}, a) = q_{[wa]}$, for any $q_{[w]} \in Q$ and any $a \in \Sigma$, where $[w]$ is an equivalence class of the relation R_L represented by a word $w \in \Sigma^*$.

The automaton M designed above accepts the language L because:
- for any $w \in L$, $\delta(q_0, w) = \delta(q_{[\varepsilon]}, w) = q_{[w]}$ (simple inductive proof justifies this evidence), i. e., $\delta(q_0, w) \in F$,
- likewise, for any $w \notin L$, $\delta(q_0, w) \notin F$.

The proof has been completed. □

8.1.5 Minimization of deterministic finite automata

The Myhill–Nerode theorem helps to minimize deterministic finite automata. First of all, note that the relation R_L induced by a regular language L is the most general equivalence relation defining a language L. Namely, an equivalence relation defines a language if and only if a language is a union of some equivalence classes of this relation. Recall, that an equivalence relation E_1 is more general than an equivalence relation E_2 if and only if equivalence classes of E_2 are included into equivalence classes of E_1.

Secondly, a deterministic automaton designed in the proof of the third implication is a minimal one concerning the number of states. If this were not true, then we would be able to build a deterministic automaton M', which has fewer states than the automaton M designed in the proof. However, the relation ρ_M considered in the proof of Myhill–Nerode theorem would have fewer equivalence classes than the relation R_L. But this is not possible due to the second implication considered in the proof of the Myhill–Nerode theorem.

Finally, an automaton designed in the proof of the Myhill–Nerode theorem is a minimal one concerning the number of states.

8.2 More grammars and automata

8.2.1 Context-free grammars versus pushdown automata

In this section, relations between context-free grammars and pushdown automata are discussed. It is shown that pushdown automata are equivalent to context-free grammars. Thus, a class of languages accepted by pushdown automata is the class of context-free languages.

Theorem 8.6. *Languages generated by context-free grammars are accepted by pushdown automata.*

Proof. Let a context-free grammar is given and the empty word is not generated in the grammar. We design a pushdown automaton accepting the language generated by this grammar. The automaton accepts with the empty stack.

We assume that a given context-free grammar

$$G = (V, T, P, S)$$

is in Greibach normal form. Let us recall that productions of a grammar in Greibach normal form are $A \to a\alpha$, where $A \in V$, $a \in T$ and $\alpha \in V^*$. For a given the word $w \in L(G)$ a leftmost derivation in G is considered. A pushdown automaton, when computes a given the word w, follows this leftmost derivation in G. A pushdown automaton equivalent to the grammar G is:

$$M = (\{q_0, q\}, T, V \cup \{\rhd\}, \delta, q_0, \rhd, \lhd, \emptyset)$$

where the transition function is designed as follows:
- begin with a given the word $w \in T^*$ on the input of M and with the initial symbol S of the grammar G on the stack,
- accept if the end-of-input symbol $\lhd$ is on the input and the initial stack symbol $\rhd$ is on the stack,
- if $a \in T$ is an input symbol, $A \in V$ is a top symbol of the stack and there is a production $A \to a\alpha$ in the grammar G, $\alpha \in V^*$, then read the input symbol and replace the top symbol of the stack with α (remove A from the stack, push on the stack symbols of α in reverse order),
- reject in all other cases.

These rules could be re-written as follows:
- $\delta(q_0, \varepsilon, \rhd) = \{(q, S \rhd)\}$,
- $\delta(q, a, A) = \{(q, \alpha) : A \to a\alpha \in P\}$,
- $\delta(q, \lhd, \rhd) = \{(q, \rhd)\}$.

Modification of the design in the case when ε is included in the language is fairly easy. The automaton should be able to pop the top symbol of the stack up in its first transition, i. e., the first rule of the presented above should be replaced with:
- $\delta(q_0, \varepsilon, \rhd) = \{(q, S \rhd), (q, \rhd)\}$.

A formal proof is based on mathematical induction concerning the length of derivation. □

Notice that an automaton designed in Theorem 8.6 is, in general, a nondeterministic one. Nondeterminism is raised by the ambiguity of grammar. If a grammar in Greibach normal form is simple, i. e., satisfies the Greibach uniqueness condition, then an automaton is a deterministic one.

Example 8.4. Design a pushdown automaton equivalent to the grammar

$$G = (\{S, A\}, \{a, b\}, \{S \to aAA, A \to a \mid aS \mid bS\}, S)$$

Provide a computation of the designed automata for the word $w = abaaaa$.

Solution. The grammar G is in Greibach normal form. An equivalent pushdown automaton is:

$$M = (\{q_0, q\}, \{a, b\}, \{S, A, \rhd\}, \delta, q_0, \rhd, \lhd, \emptyset)$$

with the transition function given in Table 8.2.

The automaton accepts with the empty stack. Recalling: if a transition function is undefined, which is denoted as the empty entry in the transition table, then a pushdown automaton rejects its input.

Table 8.2: Transition tables of the pushdown automaton equivalent to the grammar given in Example 8.4.

	a	b	ε	$\lhd$
$\delta(q_0)$				
S				
A				
$\rhd$			$(q, S\rhd)$	
$\delta(q)$				
S	(q, AA)			
A	(q, S) (q, ε)	(q, S)		
$\rhd$				ACC

A computation for the given word w is presented in Figure 8.11. Note that the word is accepted because there is a path from the root to an accepting leaf (accept with empty stack and empty input).

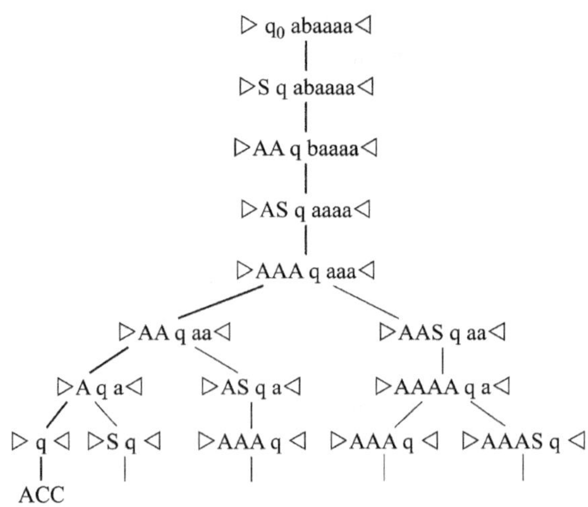

Figure 8.11: A computation for the given word "*abaaaa*".

Example 8.5. Develop a method of designing pushdown automata equivalent to $LL(1)$ grammars. Build an automaton equivalent to the grammar given in Example 11.2 of Chapter 11.

Solution. Assume that a $LL(1)$ grammar is given:

$$G = (V, T, P, S)$$

$LL(1)$ grammars satisfy a uniqueness condition, which is similar to the Greibach uniqueness condition. Therefore, it would be possible to design a deterministic automaton equivalent to a given $LL(1)$ grammar. However, productions of a $LL(1)$ grammar may have a right-hand side not beginning with a terminal symbol or may be nullable, which makes an automaton to be a nondeterministic one. In practice, if we can check an input symbol without doing an immediate transition, an automaton may be turned into a deterministic one. Below, a general, nondeterministic method for designing an automaton is given. An automaton is defined as

$$M = (\{q_0, q\}, T, V \cup T \cup \{\rhd\}, \delta, q_0, \rhd, \lhd, \emptyset)$$

where the transition function is built according to the following rules (those are slightly modified formulated in the proof of Theorem 8.6):

1. begin with a given word $w \in T^*$ at the input of M and with the initial symbol S of the grammar G on the stack,
2. accept if the end-of-input symbol $\lhd$ is at the input and the initial stack symbol $\rhd$ is on the stack,
3. if the same symbol $a \in T$ of the input alphabet is at the input and at the stack, read the input symbol and remove the top symbol of the stack,
4. if $A \in V$ is a top symbol of the stack and there is a A-production in the grammar, then choose nondeterministically an A-production $A \to \alpha$, $\alpha \in (V \cup T)^*$ and replace the top symbol of the stack with α (remove A from the stack, push on the stack symbols of α in reverse order),
5. reject in all other cases.

These rules could be re-written as follows:
- $\delta(q_0, \varepsilon, \rhd) = \{(q, S \rhd)\}$,
- $\delta(q, a, a) = \{(q, \varepsilon)\}$ for $a \in T$,
- $\delta(q, \varepsilon, A) = \{(q, a\alpha) : A \to a\alpha \in P\}$ for $A \in V \cup \{S\}$,
- $\delta(q, \lhd, \rhd) = \{(q, \rhd)\}$.

A formal proof is based on mathematical induction completed concerning the length of derivation.

An automaton equivalent to the grammar given in Example 11.2 of Chapter 11 is as follows:

$$M = (\{q_0, q\}, \Sigma, \Gamma, \delta, q_0, \rhd, \lhd, \emptyset)$$

where:
- $\Sigma = \{id, +, -, *, /, \uparrow, (,)\}$,
- $\Gamma = \{E, E_l, T, T_l, P, P_l, Q, \rhd, id, +, -, *, /, \uparrow, (,)\}$,
- δ – the transition function is given in Table 8.3. In this table, empty entries denote rejection of an input. For the sake of clarity and space saving, the values $+TE_l$, $-TE_l$, ε should be read as $\{(q, +TE_l), (q, -TE_l), (q, \varepsilon)\}$ (brackets and the state q are dropped), i. e., only symbols replacing a top symbol of the stack are shown. The symbol τ stands for any terminal symbol $\rhd, id, +, -, *, /, \uparrow, (,)$, i. e., $\delta(q, \tau, \tau) = (q, \varepsilon)$ for $\tau \in \{\rhd, id, +, -, *, /, \uparrow, (,)\}$.

This automaton can be turned into a deterministic one, assuming that an input symbol can be reused. In other words, if an input can be checked before a ε-transition is (nondeterministically) done, then a suitable transition could be (deterministically) chosen based on the uniqueness condition of $LL(1)$ grammars and SELECT sets of relevant productions.

Table 8.3: A pushdown automaton equivalent to the $LL(1)$ grammar given in Example 11.2 of Chapter 11.

δ	τ	ε
E		TE_l
E_l		$+TE_l, -TE_l, \varepsilon$
T		PT_l
T_l		$*PT_l, /PT_l, \varepsilon$
P		QP_l
P_l		$\uparrow QP_l, \varepsilon$
Q		$-Q, (E), id$
τ	ε	

Theorem 8.7. *Languages accepted by pushdown automata are generated by context-free grammars.*

Due to its complexity, proof of this theorem does not fit this book. On the other hand, this theorem has no impact on other parts of this book. Therefore, the formal proof is skipped. The reader can consult the proof in, for instance, [4].

8.2.2 Unrestricted grammars versus Turing machines

Theorem 8.8. *Languages generated by unrestricted grammars are accepted by Turing machines.*

Proof. We will describe a Turing machine, which accepts a language generated by a given unrestricted grammar. A detailed description of such a machine does not fit this book. Moreover, such a huge and detailed description would be far away from an easy understanding of how such a machine works.

Let us assume that an unrestricted grammar $G = (V, T, P, S)$ is given. We design a Turing machine, which will nondeterministically follow a derivation of a given the word. This machine has two tapes. An input word $w = a_1 a_2 \ldots a_n$ is kept on the first tape. An intermediate word of a derivation is stored on the second tape. A computation is terminated when the contents of the second tape is equal to the input word w. Note that we do not put any restriction on a derivation; it may neither be a leftmost nor a rightmost one. The machine realizes the following algorithm:

1. initialize the second tape with the initial symbol S of the grammar G, i. e., put this symbol in a cell of the tape,
2. compare contents of both tapes. If they are equal, switch to an accepting state and terminate computation,
3. nondeterministically choose a symbol in a word on the second tape (e. g. such a choice is done by the Turing machine designed in Example 5.6),
4. nondeterministically choose a production $\alpha \to \beta$ of the grammar G,
5. compare the left-hand side α of this production with the part of the word on the second tape, which begins with the symbol chosen in point 3. If they match, replace the part of the word with the right-hand side of the production (perhaps, a remaining part of the word, which follows the part α to be replaced, needs a shifting left or right, depending on whether $|\alpha| < |\beta|$ or $|\alpha| > |\beta|$),
6. go back to the step 2.

A Turing machine realizing the above algorithm accepts an input word if this word is generated in the grammar G. This is guaranteed by the fundamental assumption about nondeterminism. It is assumed that a nondeterministic choice always leads to acceptance of input if such input can be accepted, cf. Section 5.2. If an input word is not generated in the grammar G, then such a Turing machine will fall into an infinite computation.

On the other hand, if there is a sequence of transitions leading to acceptance by such a Turing machine, then this sequence of transitions defines a derivation in the grammar G.

It can be proved by induction that the content of the second tape is an intermediate word of some derivation in the grammar G. Then, based on the fundamental assumption of nondeterminism, it can be concluded that a derivation of an input word is generated in the grammar. Thus, $L(G) = L(M)$. □

Theorem 8.9. *Languages accepted by Turing machines are generated by unrestricted grammars.*

As in case of Theorem 8.7 the formal proof is skipped. The reader can consult the proof in, for instance, [4].

8.2.3 Context-sensitive grammars versus linear bounded automata

Theorem 8.10. *Languages generated by context-sensitive grammars are accepted by linear bounded automata.*

Proof. The proof is similar to the one in Theorem 8.8. Instead of using a two-tape Turing machine with tapes infinite two-way, we use a linear bounded automaton with two tapes, both of length equal to the length of an input word. This restriction may cause that it is impossible to perform an operation of the pint 5 of the proof in Theorem 8.8. The reason for such a failure is that there may be no room for replacing a part of the word with the right-hand side of a production. In such a case, an automaton rejects its input. It is worth underlining that an automaton used here is nondeterministic. According to the fundamental assumption of nondeterminism, we may interpret a computation of an automaton as a sequence of configurations correctly chosen. To summarize, a such designed linear bounded automaton accepts the language generated by a given unrestricted grammar. □

Theorem 8.11. *Languages accepted by linear bounded automata are generated by context-sensitive grammars.*

As in case of Theorem 8.7 the formal proof is skipped. The reader can consult the proof in, for instance, [4].

9 Around the hierarchy of languages

In this section, we define more operations on languages: substitutions, homomorphism and quotients. Then we prove closure properties of classes of languages with regard to operations on languages. Finally, we define the Chomsky hierarchy of families of languages and its extension and prove their properties.

9.1 More operations on languages

9.1.1 Substitutions, homomorphisms

Definition 9.1. Let Σ and Δ are alphabets. A mapping f of letters of an alphabet Σ into languages over an alphabet Δ:

$$f : \Sigma \to 2^{\Delta^*}$$

is called substitution. A substitution on an alphabet Σ can be extended

– to words over an alphabet Σ:

$$\begin{aligned} f &: \Sigma^* \to 2^{\Delta^*} \\ f(\varepsilon) &= \varepsilon \\ f(wa) &= f(w) \circ f(a) \quad (\forall w \in \Sigma^*)(\forall a \in \Sigma) \end{aligned}$$

– and to languages over an alphabet Σ:

$$\begin{aligned} f &: 2^{\Sigma^*} \to 2^{\Delta^*} \\ f(L) &= \bigcup_{w \in L} f(w) \quad (\forall L \subset \Sigma^*) \end{aligned}$$

This general definition of substitution may be restricted to a class of languages, assuming that values of substitution are languages of this class as well as arguments of a substitution are languages of this class as well. In this section discussion on substitutions is restricted to regular languages. Explicitly, we assume that substitutions are mappings: $f : \Sigma \to RgL(\Delta), f : \Sigma^* \to RgL(\Delta)$ and $f : RgL(\Sigma) \to RgL(\Delta)$, where $RgL(\Sigma)$ and $RgL(\Delta)$ are the classes of regular languages over alphabets Σ and Δ, respectively.

The definition of substitution could be reformulated to regular expressions.

Definition 9.2. Let Σ and Δ are alphabets. A mapping F of letters of an alphabet Σ into regular expressions $REx(\Delta)$ over an alphabet Δ:

$$f : \{a : a \in \Sigma\} \to REx(\Delta)$$

https://doi.org/10.1515/9783112223123-012

A substitution on an alphabet Σ can be extended to regular expressions $REx(\Sigma)$ over an alphabet Σ:

$$f : REx(\Sigma) \to REx(\Delta)$$

such that

$$\begin{aligned} f(\emptyset) &= \emptyset \\ f(\varepsilon) &= \varepsilon \\ f(a) &= \text{the value defined by the mapping } f \quad (\forall a \in \Sigma) \end{aligned}$$

and

$$\begin{aligned} f(s+t) &= (f(s)+f(t)) \\ f(s \circ t) &= (f(s) \circ f(t)) \\ f(s^*) &= ((f(s))^*) \end{aligned}$$

where $\emptyset$, ε and a for $a \in \Sigma$ are basic regular expressions, s and t are regular expressions (used in inductive step of the definition).

Remark 9.1. In the class of regular languages, substitutions can be considered alternatively with regard to languages or concerning regular expressions generating these languages.

Example 9.1. Let $\Sigma = \{a, b\}$ and $\Delta = \{0, 1\}$ are alphabets and f is a substitution:

$$\begin{aligned} f(a) &= \{00, 0000, 000000, \ldots\} \\ f(b) &= \{1, 11, 111, 1111, 11111, \ldots\} \end{aligned}$$

Redefine this substitution to regular expressions and compute its value for given regular expressions: bab and $a^*(a+b)b^*$.

Solution. The language $\{00, 0000, 000000, \ldots\}$ is generated by the regular expression $00(00)^*$, the language $\{1, 11, 111, 1111, 11111, \ldots\}$ is generated by the regular expression 11^*. Hence, the substitution reformulated to regular expressions is as follows (notice that, in light of Remark 2.2 of Chapter 2, unnecessary brackets are removed):

$$\begin{aligned} f(a) &= 00(00)^* \\ f(b) &= 11^* \end{aligned}$$

Now, let us compute values of the substitution for given regular expressions:

- $f(bab) = f(b)f(a)f(b) = (11^*)(00(00)^*)(11^*) = 11^*00(00)^*11^*$,
- $f(a^*(a+b)b^*) = f(a^*)f(a+b)f(b^*) = (f(a))^*(f(a)+f(b))(f(b))^* = (00(00)^*)^*(00(00)^* + 11^*)(11^*)^* = (00)^*(00(00)^* + 11^*)1^* = (00)^*001^* + (00)^*11^* = (00)^*(00+1)1^*$.

Definition 9.3. Let Σ and Δ are alphabets. A substitution $h : \Sigma \to 2^{\Delta^*}$, such that $|h(a)| = 1$ for all $a \in \Sigma$, is called a homomorphism. In other words, a homomorphism is such a substitution, which yields one-word languages. An extension of a homomorphism to words and languages is a relevant extension of a substitution.

Remark 9.2. A homomorphism h is identified with a mapping $h : \Sigma \to \Delta^*$, which yields a word over the alphabet Δ for a letter of the alphabet Σ rather than a language including only this word.

Definition 9.4. Let $h : \Sigma \to 2^{\Delta^*}$ is a homomorphism. An inverse homomorphic image of a word $w \in \Delta^*$ is a set of words (language):

$$h^{-1}(w) = \{x \in \Sigma : h(x) = w\}$$

An inverse homomorphic image of a language $L \subset \Delta^*$ is a set of words:

$$h^{-1}(L) = \bigcup_{w \in L} h^{-1}(w) = \{x \in \Sigma^* : h(x) \in L\}$$

Example 9.2. Let $\Sigma = \{a, b\}$ and $\Delta = \{0, 1\}$ are alphabets and h is a homomorphism:

$$h(a) = 010$$
$$h(b) = 101$$

Find an inverse homomorphic image of a language L generated by the regular expression $(\varepsilon + 0)(10)^*(\varepsilon + 1)$.

Solution. Notice that the language L includes words with alternating 0's and 1's, i. e., none a word include two successive 0's or 1's:

$$L = \{\varepsilon, 0, 1, 01, 10, 010, 101, 0101, 1010, 01010, 10101, \ldots\}$$

The language L includes the empty word ε of length 0 and two words of any length greater than 0. Of course, only words of length divisible by 3 can be yielded by the homomorphism h. This means that the inverse homomorphic image h^{-1} yields the empty set for any word of length not divisible by 3. Having this in mind, we can show that the inverse homomorphic image works for successive words of L of length divisible by 3:
- $h^{-1}(\varepsilon) = \{\varepsilon\}$,
- $h^{-1}(010) = \{a\}, h^{-1}(101) = \{b\}$,
- $h^{-1}(010101) = \{ab\}, h^{-1}(101010) = \{ba\}$,
- $h^{-1}(010101010) = \{aba\}, h^{-1}(101010101) = \{bab\}$,
- $h^{-1}(010101010101) = \{abab\}, h^{-1}(101010101010) = \{baba\}$,
- ....

As it is seen, the inverse homomorphic image $h^{-1}(L)$ includes words with alternating a's and b's, i. e., the same language as L with 0's replaced by a's and 1's replaced by b's. The language $h^{-1}(L)$ is, of course, generated by the same regular expression, which generates L with the same replacements: $(\varepsilon + a)(ba)^*(\varepsilon + b)$. Despite that $h^{-1}(L) \simeq L$, it is not true that $h(h^{-1}(L)) \simeq L$. The homomorphic image of the language generated by the regular expression $(\varepsilon + a)(ba)^*(\varepsilon + b)$, i. e., having all words with alternating a's and b's, includes words with alternating 0's and 1's of length divisible by 3.

9.1.2 Quotients

Quotients of languages are actually applied to words. Quotient of words is essentially the opposite of concatenation. A quotient of words is a kind of reduction of one word, a dividend, by another one, a divisor. Two types of quotients are defined: the right quotient and the left quotient.

Definition 9.5. Let L_1 and L_2 are languages over an alphabet Σ.

The right quotient of a language L_1 by a language L_2 is the language L_1/L_2 consisting of such words over the alphabet Σ, which concatenated with words of the divisor give words of the dividend:

$$L_1/L_2 = \{x \in \Sigma^* : (\exists y \in L_2) xy \in L_1\}$$

The left quotient of a language L_1 by a language L_2 is the language $L_1\backslash L_2$ consisting of such words over the alphabet Σ words, which concatenated to words of the divisor give words of the dividend:

$$L_1\backslash L_2 = \{y \in \Sigma^* : (\exists x \in L_2) xy \in L_1\}$$

Example 9.3. Let $L_1 = \{a^n b^n c^n : n \geq 0\}$ and $L_2 = \{a^k b^l c^m : k, l, m \geq 0\}$. Find quotients: L_1/L_2, $L_1\backslash L_2$, L_2/L_1 and $L_2\backslash L_1$.

Solution. Let us consider every case:

1. L_1/L_2 – analyzing definition of the right quotient we get:

$$L_1/L_2 = \{w \in \{a, b, c\}^* : w \circ a^k b^l c^m = a^n b^n c^n,\ k, l, m, n \geq 0\}$$

The following cases are possible with regard to w:
- $w = a^n b^n c^n$ and $a^k b^l c^m = \varepsilon$, i. e. $k = l = m = 0$ for $n > 0$ (notice that the case $n = 0$ is considered in the forth item of this list) or
- $w = a^n b^n c^p$ and $a^k b^l c^m = c^{n-p}$ for $n > p \geq 0$ or
- $w = a^n b^p$ and $a^k b^l c^m = b^{n-p} c^n$ for $n > p \geq 0$ or
- $w = a^p$ and $a^k b^l c^m = a^{n-p} b^n c^n$ for $n > p \geq 0$.

Finally:
$L_1/L_2 = L_1 \cup \{a^n b^n c^p : n > p \geq 0\} \cup \{a^n b^p : n > p \geq 0\} \cup \{a^p : p \geq 0\}$.

2. $L_1 \backslash L_2$ – solution is similar to the solution of the above case:
$L_1 \backslash L_2 = \{w \in \{a, b, c\}^* : a^k b^l c^m \circ w = a^n b^n c^n, k, l, m, n \geq 0\}$,
$L_1 \backslash L_2 = L_1 \cup \{a^p b^n c^n : n > p \geq 0\} \cup \{b^p c^n : n > p \geq 0\} \cup \{c^p : p \geq 0\}$.
3. L_2/L_1 and $L_2 \backslash L_1$ give the same result. Let us consider L_2/L_1:
$L_2/L_1 = \{w \in \{a, b, c\}^* : w \circ a^n b^n c^n = a^k b^l c^m,\ k, l, m, n \geq 0\}$
Notice that an empty word ε is in the divisor. So, any word $w = a^k b^l c^m$ is in the quotient. On the other hand, only a string of a's concatenated with a nonempty divisor's word gives a word of the dividend. Analysis of $L_2 \backslash L_1$ is similar.
Finally: $L_2/L_1 = L_2 \backslash L_1 = L_1$.

Remark 9.3. The definition of quotients of languages could be reformulated to regular expressions instead of languages. Such a formulation corresponds, of course, to quotients of languages generated by regular expressions, i. e., to quotients of regular languages.

Example 9.4. Languages R_1, R_2 and R_3 are generated by regular expressions $r_1 = 0^*10^*$, $r_2 = 10^*1$ and $r_3 = 0^*1$. Find the following quotients R_1/R_2, $R_1 \backslash R_2$, R_1/R_3, $R_1 \backslash R_3$, R_2/R_3, $R_2 \backslash R_3$.

Solution. Results are given in a form of regular expressions, thus a symbol $\simeq$ is used instead of equality – $R_1/R_2 \simeq \emptyset, R_1 \backslash R_2 \simeq \emptyset, R_1/R_3 \simeq 0^*, R_1 \backslash R_3 \simeq 0^*, R_2/R_3 \simeq 10^*, R_2 \backslash R_3 \simeq 0^*1$.

9.1.3 Building automata with quotients

A language $L = L(M)$ accepted by a deterministic finite automaton $M = (Q, \Sigma, \delta, q_0, F)$ is a set of words $L = \{w \in \Sigma^* : \delta(q_0, w) \in F\}$. Let consider languages:

- $L_a = \{w \in \Sigma^* : \delta(q_a, w) \in F\}$, where $\delta(q_0, a) = q_a$ for $a \in \Sigma$. These languages are accepted by deterministic automata $M = (Q, \Sigma, \delta, q_a, F)$. Note that L_a is derived from L by removing the first letter from words of L, i. e., $L_a = \{u \in \Sigma^* : (\exists w \in L)\ au = w\}$. The last formula defines the left quotient of the language L with the divisor language $\{a\}$ (the divisor language includes one word of unit length). i. e., $L_a = L \backslash \{a\}$,
- $L_{ab} = \{w \in \Sigma^* : \delta(q_{ab}, w) \in F\}$, where $q_{ab} = \delta(q_0, ab)$ for $a, b \in \Sigma$. L_{ab} is derived from L by removing two leading first letters from words of L or – in other words – L_b is derived from L_a by removing the first letter from words of it: $L_{ab} = L \backslash \{ab\} = L_a \backslash \{b\}$,
- $L_{abc} = \{w \in \Sigma^* : \delta(q_{abc}, w) \in F\}, \delta(q_0, a) = q_{ab}$ for $a, b, c \in \Sigma^*$,
- ....

How many languages do we get in the above process of quotients? As many as words in the language L, at a glance. Nevertheless, since languages are tied to states of a finite automaton, we get no more languages than the number of states. On the other hand, languages are computed as quotients by words over the alphabet of this language. They may be tied to any deterministic finite automaton, including an automaton with a minimal number of states. Hence, the number of languages does not depend on a particular automaton. From the Myhill–Nerode theorem, we conclude that these languages are tied with equivalence classes of the relation R_L induced by a regular language L rather than a particular deterministic finite automaton accepting L. However, these languages are not equivalence classes of R_L. Moreover, they are not equivalence classes of any equivalence relation.

Now, consider which of the above languages correspond to equivalence classes of R_L included in the language L. Take a particular language L_u. Note that there are many $u \in \Sigma^*$ defining the same language. In fact, a set of words defining a particular language can be written as $E_w = \{u \in \Sigma^* : L_u = L_w\}$. If an automaton M is a minimal one, then $E_w = \{u \in \Sigma^* : \delta(q_0, u) = \delta(q_0, w)\}$ is an equivalence class of R_L, cf. the Myhill–Nerode theorem. If v is a shortest word in E_w, then the state $\delta(q_0, v) = \delta(q_0, w)$ is accepting one if and only if $v \in L$, what is equivalent to $\varepsilon \in L_w$.

As a conclusion of this discussion, we get the following proposition:

Proposition 9.1. *A regular language L over an alphabet Σ^* is accepted by a deterministic finite automaton*

$$M = (Q, \Sigma, \delta, q_0, F)$$

where:

- $Q = \{q_{L_w} : L_w = L\backslash\{w\} \wedge w \in \Sigma^*\}$, *i. e., states are labeled by quotients of languages,*
- $q_0 = q_L$,
- $F = \{q_{L_w} : \varepsilon \in L_w\}$,
- $\delta(q_{L_w}, a) = q_{L_{wa}}$ *for* $a \in \Sigma$, $w \in \Sigma^*$ *and, in this convention,* $L_\varepsilon = L$.

Moreover, this automaton is a minimal one (with regard to a number of states) accepting L assuming that equality of any two languages obtained from different quotients is identified.

Example 9.5. Let a language L be generated by the regular expression $r = (aa)^*(aa + bb)(bb)^*$. Employ Proposition 9.1 to design a deterministic finite automaton accepting L.

Solution. Calculating quotients let us use regular expressions instead of languages, one has:

- $r = (aa)^*(aa + bb)(bb)^*$,
- $r_a = r\backslash a = a(aa)^*(aa + bb)(bb)^* + a(bb)^*$,
- $r_b = r\backslash b = b(bb)^*$,
- $r_{aa} = r_a\backslash a = (aa)^*(aa + bb)(bb)^* + (bb)^*$,
- $r_{ab} = r_a\backslash b = \emptyset$,
- $r_{ba} = r_b\backslash a = \emptyset$,
- $r_{bb} = r_b\backslash b = (bb)^*$,

- $r_{aaa} = r_{aa} \backslash a = r \backslash a = r_a$,
- $r_{aab} = r_{aa} \backslash b = b(bb)^* + b(bb)^* = r_b$,
- $r_{aba} = r_{abb} = r_{baa} = r_{bab} = \emptyset$,
- $r_{bba} = r_{bb} \backslash a = \emptyset$,
- $r_{bbb} = r_{bb} \backslash b = b(bb)^* = r_b$.

No new language can be yielded when quoting is continued.

The automaton is $M = (\{r, r_a, r_b, r_{aa}, r_{bb}, \emptyset\}, \{a, b\}, \delta, r, \{r_{aa}, r_{bb}\})$, where the transition function is given as the transition diagram in Figure 9.1.

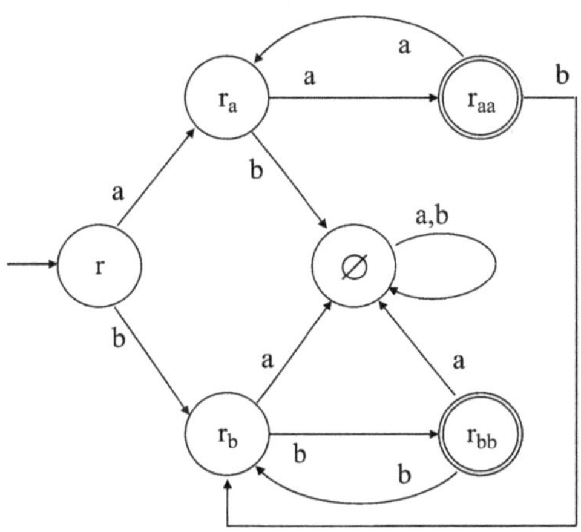

Figure 9.1: The transition diagram of the automaton designed in Example 9.5.

9.2 Closure

In this section, we investigate the closure of classes of languages with regard to operations on languages. Operations are defined in Section 1.5.2 of Chapter 1.1 and above in this Chapter. We consider classes of languages examined in the previous sections: regular, context-free, context-sensitive, recursive, recursively enumerable classes of languages, i. e., *RgL*, *CFL*, *CSL*, *RkL*, *REL* classes, and the class of all languages, the *ALL* class.

Let us examine this problem from two points of view. The first attempt employs grammars, which generate languages of relevant classes. This attempt allows showing that union is an internal operation in all classes of languages generated by grammars, i. e., in *RgL*, *CFL*, *CSL*, *REL* classes of languages. The second attempt is based on automata. It allows for proving union to be an internal operation in *RgL*, *CFL*, *CSL*, *RkL* and *REL* classes. We give proofs for both attempts, despite that in this way proofs are duplicated for some classes. We think that these proofs play utilitarian roles for discussion on closure as well as they provide useful techniques, which could be applied in solving other problems.

Remark 9.4. Taking a subset of a given set, in this case a subset of a language, is the very basic operation. However, it is not an internal operation in any class of languages, except the *ALL* class. It means that a subset of a language of any class may not belong to this class, besides the *ALL* class.

Remark 9.5. The *ALL* class is closed with regard to any operation. Therefore, the further discussion does not take this class into account.

Proposition 9.2. *Union is an internal operation in all classes of languages.*

Proof. Let us assume that given two languages L_1 and L_2 belong to a given class, i. e., and they are either regular or context-free or context-sensitive or recursively enumerable. Hence, we can assume that there are grammars $G_1 = (V_1, T_1, P_1, S_1)$ and $G_2 = (V_2, T_2, P_2, S_2)$ such that $L_1 = L(G_1)$ and $L_2 = L(G_2)$, i. e., these grammars generate given languages. Of course, both grammars are in a class to which these languages belong. We can assume that $V_1 \cap V_2 = \emptyset$. Otherwise, the names of nonterminal symbols of one of these grammars should be changed.

The following grammar generates union of both languages $L = L_1 \cup L_2$, i. e., $L = L(G)$

$$G = (V_1 \cup V_2 \cup \{S\}, T_1 \cup T_2, P_1 \cup P_2 \cup \{S \to S_1 \mid S_2\}, S)$$

Note that a form of productions of a given grammar decides to which class this grammar belongs to. The grammar G belongs to the same class, to which grammars G_1 and G_2 do. This is because the new productions $S \to S_1 \mid S_2$ hold assumptions on productions of any grammar: regular, context-free, context-sensitive, or unrestricted one. Because G is in the same class, in which G_1 and G_2 are, then L belongs to the same class, to which L_1 and L_2 do.

Also note that $L = L_1 \cup L_2 \subset L(G)$, i. e., any word w of $L = L_1 \cup L_2$ is generated in G, what means that $w \in L(G)$. If $w \in L = L_1 \cup L_2$, then $w \in L_1$ or $w \in L_2$. If $w \in L_1$, then its derivation in G is of a form $S \to S_1 \to^* w$, where $S_1 \to^* w$, where is a derivation of w in G_1. A similar derivation can be built for a word $w \in G_2$. On the other hand $L \subset L = L_1 \cup L_2$. If a derivation in G is given, then it either begins with the production $S \to S_1 \to \cdots$ or with the production $S \to S_2 \to \cdots$. A remaining part of a derivation is either a derivation in G_1 or in G_2.

Hence, we have proved that the grammar G and the language $L = L(G)$ belong to the same class to which grammars G_1 and G_2 and languages L_1 and L_2 do. Then we showed that the grammar G generates words of $L = L_1 \cup L_2$ and oppositely. This completes the proof *based on grammars* that union is an internal operation in classes given above.

Consequently, we have another prove that union is an internal operation in all classes of languages. This proof is *based on automata.* Since we do not have a class of grammars generating recursive languages, this is the only proof that union is an internal operation in the *RkL* class.

For the context-sensitive, recursive or recursively enumerable languages L_1 and L_2 we can assume that we have linear bounded automata, Turing machines with the stop property or Turing machines M_1 and M_2, which accept these languages, i. e. $L = L(M_1)$ and $L = L(M_2)$, respectively. In Chapter 5 we have built a Turing machine, which accepts union of L_1 and L_2; namely, in Problem 5.6 we designed a Turing machine accepting $L(M_1) \cup L(M_2)$, for given Turing machines M_1 and M_2. Similarly, Turing machines with the stop property and linear bounded automata are deemed in Problem 5.10 and Problem 5.11. Likewise, a push-down automaton is designed in Problem 6.6 in Chapter 6. A construction of finite automaton is given in Problem 7.4 in Chapter 7. □

Proposition 9.3. *Concatenation is an internal operation in all classes of languages.*

Proof. This proof is quite similar to the proof of Proposition 9.2. The difference is in a construction of the set of productions of a grammar generating concatenation of languages. The productions $S \rightarrow S_1 \mid S_2$ should be replaced with the production $S \rightarrow S_1S_2$. This task is also discussed in Problem 5.8, Problem 5.10 and Problem 5.11 in Chapter 5 and in Problem 7.4 in Chapter 7. □

Proposition 9.4. *Kleene closure is an internal operation in all classes of languages.*

Proof. Again, this proof is quite similar to the proof of Proposition 9.2. Suppose that a grammar $G_1 = (V_1, T_1, P_1, S_1)$ generates a language L_1. Then the grammar $G = (V_1, T_1, P_1 \cup \{S_1 \rightarrow \varepsilon \mid S_1S_1\})$ generates $L = (L_1)^*$. This task is also solved in Problem 5.9, Problem 5.10 and Problem 5.11 in Chapter 5 and in Problem 7.4 in Chapter 7. □

Proposition 9.5. *Substitution is an internal operation in classes of languages RgL, CFL, CSL, REL and ALL.*

Proof. We prove closure of the class of context-free languages with regard to substitution based on the class of context-free grammars. Notice that each class *RgL, CFL, CSL* and *REL* of languages is generated by a corresponding class of grammars. Therefore, the proof can be repeated for other classes of languages with the replacement of the class of grammars.

Let us assume that Σ and Δ are alphabets and $f : \Sigma \rightarrow 2^{\Delta^*}$ a substitution such that $f(a)$ is a context-free language $L_a \subset \Delta^*$ for each $a \in \Sigma$ and that $L \subset \Sigma^*$ is a context-free language. Let us assume that the context-free grammars $G = (V_a, \Delta, P_a, S_a)$ generate languages L_a for each $a \in \Sigma$ and that $G = (V, \Sigma, P, S)$ is a context-free grammar generating the language L. Based on these assumptions we will design a context-free grammar generating the language $L_f = f(L)$, what ends the proof that the class of context-free languages is closed under substitution.

We design a context-free grammar $G_f = (V_f, \Delta, P_f, S)$ which generates the language L_f. Let us assume that sets of nonterminal symbols V_a, $a \in \Sigma$ are pairwise disjoint and disjoint with the set V. Then:

- $V_f = V \cup \bigcup_{a\in\Sigma} V_a$,

- $P_f = P' \cup \bigcup_{a \in \Sigma} P_a$
  where P' are productions from P:
  - unchanged, if they do not contain any terminal,
  - each terminal symbol $a \in \Sigma$ is replaced with S_a, if they contain terminals.

Justification is straightforward: productions from P' generate words of the language L, but in these words, we have initial symbols S_a of grammars G_a instead of terminals $a \in \Sigma$. Then, each such symbol S_a is used to generate a word of the language L_a. This is consistent with Definition 9.1. □

Proposition 9.6. *Homomorphism is an internal operation in classes of languages RgL, CFL, CSL, REL and ALL.*

Proof. Homomorphism is a special case of substitution. Proof is a direct consequence of proof for substitution. □

Proposition 9.7. *Intersection is an internal operation in RgL, CSL, RkL and REL classes of languages.*

Proof. Compare solutions of Problem 7.4 in Chapter 7 and Problem 5.6 and Problem 5.11 in Chapter 5. □

Proposition 9.8. *Intersection is not an internal operation in CFL class of languages.*

Proof. Consider languages $L_1 = \{w \in \{a,b,c\}^* : w = a^k b^k c^l,\ k,l > 0\}$ and $L_2 = \{w \in \{a,b,c\}^* : w = a^k b^l c^l,\ k,l > 0\}$. Both are context-free, but intersection of them $L = \{w \in \{a,b,c\}^* : w = a^k b^k c^k,\ k > 0\}$ is not context free. □

Proposition 9.9. *Complement is an internal operation in RgL, CSL and RkL classes of languages.*

Proof. Compare solutions of Problem 7.4 in Chapter 7 and Problem 5.7 and Problem 5.11 in Chapter 5. □

Proposition 9.10. *Complement is not an internal operation in CFL class.*

Proof. Compare solution of Problems 3.6 and 3.7 in Chapter 3. □

Proposition 9.11. *Complement is not an internal operation in REL class.*

Proof. The complement of the diagonal language is recursively enumerable, but its complement (i. e., the diagonal language) is not, compare Example 9.6 and Propositions 9.17 and 9.18. Also, the universal language defined in 9.7 is recursively enumerable, but its complement is not. □

Proposition 9.12. *Quotients (left and right) are internal operations in the class of regular languages.*

Proof. Let us assume that we have a regular language $L_1 \subset \Sigma^*$ accepted by a finite automaton $A = (Q, \Sigma, \delta, q_0, F)$ and a language $L_2 \subset \Sigma^*$ (not necessarily regular). We will design automata accepting the right quotients L_1/L_2 and the left quotient $L_1\backslash L_2$.

The right quotient
L_1/L_2 is accepted by the finite automaton

$$A_r = (Q, \Sigma, \delta, q_0, F_r)$$

where

$$F_r = \{q \in Q : (\exists y \in L_2)\delta(q, y) \in F\}$$

Indeed, for any $x \in \Sigma^*$, $\delta(q_0, x) \in F_r$ if and only if there exists $y \in L_2$ such that $\delta(q_0, xy) = \delta(\delta(q_0, x), y) \in F)$.

The left quotient
$L_1\backslash L_2$ is accepted by the finite automaton

$$A_l = (Q \cup \{q_l^0\}, \Sigma, \delta_l, q_l^0, F)$$

where

$$\begin{aligned} \delta_l(q, a) &= \{\delta(q, a)\} \quad \text{for all } q \in Q,\ a \in \Sigma \\ \delta_l(q_l^0, \varepsilon) &= \{q \in Q : (\exists x \in L_2)\delta(q_0, x) \in F\} = \bigcup_{x \in L_2} \{\delta(q_0, x)\} \\ \delta_l(q, u) &= \emptyset \quad \text{otherwise} \end{aligned}$$

Indeed, for any $y \in \Sigma^*$, $\delta_l(q_l^0, x) \subset F$ if and only if there exists $x \in L_2$ such that $\delta(q_0, xy) = \delta(\delta(q_0, x), y) \in F)$.

The above construction is not efficient: it is necessary to check all words of a language L_2 in order to complete automata construction. If we assume that L_2 is a regular language, we can effectively find automata accepting quotients L_1/L_2 and $L_1\backslash L_2$ checking words of L_2 shorter than the constant in pumping lemma. This is a direct consequence of pumping lemma, which says that computation for any word not shorter than the constant in this lemma can be shortened without changing its last state, cf. proof of pumping lemma. □

9.3 The hierarchy of languages

Proposition 9.13. *The class RgL of regular languages in included, but not equal to the class CFL of context-free languages.*

Table 9.1: Closure with regard to operations on languages.

is internal in	*RgL*	*CFL*	*CSL*	*RkL*	*REL*	*ALL*
union	+	+	+	+	+	+
concatenation	+	+	+	+	+	+
Kleene closure	+	+	+	+	+	+
substitution	+	+	+		+	+
homomorphism	+	+	+		+	+
intersection	+	–	+	+	+	+
complement	+	–	+	+	–	+
quotients	+					+

Proof. According to Theorem 8.1 and Theorem 8.3 regular languages are generated by right-linear grammars. On the other hand, right-linear grammars are context-free ones. Thus, we have inclusion. Moreover, the language $L = \{w \in \{a, b\}^* : \#_a w = \#_b w > 0\}$ is not a regular one, but it is a context-free one, cf. Example 2.7 and Example 2.7 in Chapter 2 and Problem 3.1 in Chapter 3. □

Proposition 9.14. *The class CFL of context-free languages in included, but not equal to the class CSL of context-sensitive languages.*

Proof. Again, context-free languages are generated by context-free grammars. On the other hand, context-free grammars are also context-sensitive ones, which generate context-sensitive languages. Moreover, the language $L = \{w \in \{a, b, c\}^* : \#_a w = \#_b w = \#_c w > 0\}$ is not a context-free one, but it is a context-sensitive one, cf. Problem 3.3 in Chapter 3 and Example 4.1 in Chapter 4. Thus, we have inclusion, but not equality. □

Lemma 9.1. *There is a Turing machine with the stop property (an algorithm) to check if a given word $z = a_1 a_2 \dots a_n$ is generated by a given context-sensitive grammar $G = (V, T, P, S)$.*

Proof. The Turing machine realizes a shortest paths algorithm. Let us build a graph with nodes labeled by all words $w \in (V \cup T)^*$, which are not longer than n. Note that the initial symbol of the grammar S and the word z are among labels of nodes. There is a directed edge between two nodes labeled with words u and v if and only if there is a direct derivation of v from u in G. Checking if the word v is directly derived from u is easy and could be performed by a Turing machine with stop property. In this way, $w \in L(G)$ if and only if there is a path from the node labeled S to the node labeled w. A Turing machine with stop property could be built, which checks the existence of such a path and find the path. Such a machine implements an algorithm for path searching in a graph. Algorithms for path searching in graphs can be found in, e. g., [17, 21]. □

Lemma 9.2. *The set of context-sensitive grammars is countable, i. e., context-sensitive grammars can be enumerated with natural numbers.*

Proof. In fact, we can encode context-sensitive grammars as natural numbers. Assume that $G = (V, T, P, S)$ is a context sensitive grammar. Then the following is done:

- terminal and nonterminal symbols are replaced with binary numbers represented by a block of digits of fixed lengths. The number of binary digits necessary for encoding terminal and nonterminal symbols and an additional symbol is equal to $p = \lfloor \log_2(|V| + |T| + 1) \rfloor + 1$. Assume that the number $2^p - 1$ enumerates a special symbol, a separator. It is represented as the block $111 \dots 11$ of p binary digits 1,
- nonterminal symbols are enumerated by successive natural numbers starting with 0 (represented as the block $000 \dots 00$ of p binary digits 0). Assume that the beginning symbol S of a grammar is enumerated with 0. Of course, every number is represented by a string of p binary digits, some or all of them with non-significant zeros,
- enumeration of terminal symbols is continued with successive natural numbers following enumeration of nonterminal symbols,
- productions are represented as sequences of p-digits blocks enumerating symbols. The special symbol (i. e., the block of p ones) separates both hand sides of productions,
- a grammar is encoded as the following sequence of blocks of p binary digits:
  - encoding begins with two separators (two blocks of 1s),
  - nonterminal symbols ($|V|$ blocks of binary digits, the first one is the block of 0s),
  - the separator,
  - terminal symbols ($|T|$ blocks of binary digits),
  - the separator and a production, these two elements are repeated for every production,
  - encoding ends with two separators.

As a result, we get a binary number that encodes the given grammar G.

Note that at most one context-sensitive grammar can be encoded as a given natural number and not every number encodes a grammar, i. e., such numbers, for which binary representation is not valid code of any grammar. However, we can assume that numbers, which are not valid codes of context-sensitive grammars, encode a grammar generating the empty language. Likewise, not every natural number represents a word over the set of terminal symbols T, for instance, binary words shorter than p. Nevertheless, we can treat such natural numbers as not generated by the grammar.

This encoding is ambiguous, i. e., a given grammar may have many codes. For instance, an order of symbols or productions affects the result of encoding. All context-sensitive grammars are encoded as natural numbers and no number is a code of two grammars. Therefore, grammars can be ordered according to the smallest codes, which gives a method of enumeration of grammars and accessing the grammar encoded as a given number. Simply take binary representation of successive natural numbers and then check if it is correctly encoded grammar. If a grammar of a given code is searched, continue this process until this code is found. Of course, any grammar can be identified in this way. □

Proposition 9.15. *The class CSL of context-sensitive languages is included, but not equal to the class RkL of recursive languages.*

Proof. Context-sensitive languages are accepted by linear bounded automata. Linear bounded automata are restricted Turing machines with the stop property. Recursive languages are accepted by Turing machines with the stop property. Thus, we have the inclusion of the class CSL in the class RkL.

Now we design a language that is in RkL class, but not in CSL class. Let us build a relation $r \subset N \times N$. A pair (k, l) of natural numbers belongs to this relation if and only if the context-sensitive grammar encoded as l generates the binary word at k-th place in the canonical order, i. e., $r_{k,l} = 1$ if the l-th grammar generates the k-th word, $r_{k,l} = 0$ otherwise. Consider the language of words, which are not generated by the corresponding grammar of the code equal to index of the word, i. e., with 0 at the main diagonal in Table 9.2. This is so called diagonal language $L_d = \{w \in \{0,1\}^* : w = w_i \wedge r_{i,i} = 0\}$ in the class CSL. We will come to contrary, if we assume that L_d is context-sensitive. If it is context-sensitive, then – due to Lemma 9.2 – it is generated by a context-sensitive grammar encoded as some natural number, say number k and denote it G_k. Consider the word w_k in canonical order. If it is in L_d, then $r_{k,k} = 0$ by definition of L_d. However, $r_{k,k} = 0$ means that G_k does not generate w_k, though it should. On the other hand, if w_k does not belong to L_d, then $r_{k,k} = 1$ by definition of L_d. However, $r_{k,k} = 1$ means that G_k generates w_k, though it should not. Thus, the diagonal language L_d in the class CSL cannot be a context-sensitive one.

The language L_d is accepted by a Turing machine with the stop property. Such a machine realizes the following algorithm:
- it finds the number k of a given binary word in canonical order, i. e., $w = w_k$,
- it finds the context-sensitive grammar G_k encoded as the number k,
- it checks, if the grammar G_k generates the word $w = w_k$ or not. The method shown in Lemma 9.1 can be employed for checking.

This method allows answering the question if any word is generated by a given grammar or not. This shows that the diagonal language L_d in the class CSL belongs to the RkL class. In this way, we have proved that the CSL class is included but not equal to the RkL class. □

Lemma 9.3. *The set of Turing machines is countable, i. e., Turing machines can be enumerated with natural numbers.*

Proof. The proof is similar to the proof of Lemma 9.2. We encode Turing machines as natural numbers. Assume that $M = (Q, \Sigma, \Gamma, \delta, q_0, B, \{q_A\})$ is a Turing machine with a halting accepting state. Then we encode the machine M as a binary number in the following way:
- states, symbols of the tape alphabet Γ, symbols of the input alphabet Σ and two symbols of directions of the head shift are replaced with binary numbers represented

Table 9.2: The membership table for context-sensitive grammars.

$\delta(q_0, w)$	0	1	2	3	...	k	...
$w_0 = \varepsilon$	$r_{0,0}$	$r_{0,1}$	$r_{0,2}$	$r_{0,3}$		$r_{0,k}$	
$w_1 = 0$	$r_{1,0}$	$r_{1,1}$	$r_{1,2}$	$r_{1,3}$		$r_{1,k}$	
$w_2 = 1$	$r_{1,0}$	$r_{1,1}$	$r_{1,2}$	$r_{1,3}$		$r_{1,k}$	
$w_3 = 00$	$r_{1,0}$	$r_{1,1}$	$r_{1,2}$	$r_{1,3}$		$r_{1,k}$	
...							
w_k	$r_{k,0}$	$r_{k,1}$	$r_{k,2}$	$r_{k,3}$		?	
...							

by blocks of digits of fixed length. An additional symbol, a separator, is included in the set of codes. The number of binary digits necessary for such an encoding is equal to $p = \lfloor \log_2(|Q|+|\Sigma|+|\Gamma|+3) \rfloor + 1$. Assume that the number $2^p - 1$ enumerates a special symbol, a separator. It is represented as the string $111\dots11$ of p binary digits 1,
- states are enumerated by successive natural numbers starting with 0. Assume that the initial state is enumerated with the number 0 and the accepting state – with the number 1 (of course, every number is represented by a block of p binary digits, some or all of them with non-significant zeros),
- enumeration of symbols of Σ is continued with successive natural numbers following enumeration of states,
- then enumeration of symbols of Γ is continued with successive natural numbers following enumeration of symbols of Σ,
- then enumeration of directions of the head shift is done with the next two successive natural numbers following enumeration of Γ,
- every transition $\delta(q, X) = (p, Y, D)$ is represented as five p-digits blocks, namely these blocks represent arguments of the transition function (a state q and a tape symbol X) and result (a state p, a tape symbol Y and a direction D,
- a Turing machine is encoded as the following sequence of blocks of p binary digits:
  - the encoding begins with two separators (two blocks of 1s),
  - states ($|Q|$ blocks of binary digits, the first one is the block of 0s), recall that the initial state is encoded with the number 0, the accepting state is encoded – with the number 1,
  - the separator,
  - symbols of Σ, ($|\Sigma|$ blocks of binary digits),
  - the separator,
  - symbols of Γ, ($|\Gamma|$ blocks of binary digits),
  - the separator,
  - directions of the head moves (2 blocks of binary digits),
  - the separator,
  - the separator and a transition, these two elements are repeated for every transition (every entry of the transition table),

– the encoding ends with two separators.

This encoding is ambiguous, as a similar encoding in Lemma 9.2. Natural numbers, which are not valid codes of a Turing machine, are assumed to be codes of a machine falling in infinite computation for every input. Binary words not representing any word over Σ are assumed not to be accepted by a Turing machine. Finally, we can conclude that all Turing machines can be enumerated with natural numbers. □

Proposition 9.16. *There are languages, which are not recursively enumerable.*

Proof. The set of all words Σ^* over a given alphabet Σ in infinite and countable, cf. Example 1.2 in Chapter 1.1. The class of languages $ALL = \{L : L \subset \Sigma^*\}$ is the power set of Σ^*, so then it is uncountable. On the other hand, languages of the *REL* class are accepted by Turing machines. The set of Turing machines is countable, cf. Lemma 9.3, so the class *REL* of languages is countable. Since the class ALL is an uncountable set, then it cannot be equal to its countable subset. This proves that there are languages, which are not recursively enumerable. □

Example 9.6. Let design *the diagonal language L_d* in the class of recursively enumerable languages. Let $r \subset N \times N$ is a relation build in a similar way as in Proposition 9.15, i. e., a pair (k, l) of natural numbers belongs to this relation if and only if the Turing machine encoded as l accepts the binary word at k-th place in the canonical order. In other words, $r_{k,l} = 1$ if the l-th Turing machine accepts the k-th word, $r_{k,l} = 0$ otherwise. Note, that $r_{k,l} = 0$ means that the l-th Turing machine either terminates computation and rejects its input or it is doing infinite computation for the k-th word. The diagonal language is defined by the formula $L_d = \{w \in \{0,1\}^* : w = w_i \wedge r_{i,i} = 0\}$, i.e it includes words, for which there is zero at the main diagonal of the relation r.

Proposition 9.17. *The diagonal language L_d is not recursively enumerable.*

Proof. If we assume that $L_d \in REL$, i. e., it is recursively enumerable, then we will come to contrary. If L_d is recursively enumerable, then there exists a Turing machine, which accepts it, say the k-th machine M_k. Consider $r_{k,k}$. If $r_{k,k} = 1$, then w_k is accepted by M_k, so $w_k \in L_d$, but it should not due to definition of L_d. If $r_{k,k} = 0$, then w_k is not accepted by M_k, so $w_k \notin K_d$, but it should due to definition of L_d. Thus, the diagonal language L_d cannot be a recursively enumerable one. □

Proposition 9.18. *Complement of the diagonal language $\overline{L_d} = \{w \in \{0,1\}^* : w = w_i \wedge r_{i,i} = 1\}$ is recursively enumerable.*

Proof. In order to prove this proposition, we need to design a Turing machine accepting $\overline{L_d}$, i. e., such a Turing machine which for a given the word $w \in \{0,1\}^*$ stops computation in accepting state if and only if $w \in \overline{L_d}$. Let us recall that for a word $w \notin \overline{L_d}$ it either stops computation in not accepting state or is doing infinite computation. For a word $w \in \{0,1\}^*$, such a machine carries out the following computations:

- it encounters index i of w in canonical order, i. e., $w = w_i$,
- it finds the Turing machine M_i encoded as i,
- it simulates computation of M_i for w_i,
- it accepts $w = w_i$ if and only if M_i accepts w_i.

Notice that for any $w \in \{0,1\}^*$, for which we have one at the main diagonal of the relation r defined in Example 9.6, this Turing machine will finish computation in an accepting state and only for such words over the alphabet {0,1}. □

Example 9.7. *The universal language $L_u \subset \{0,1\}^*$ includes words $\langle M\ w\rangle$ being concatenation of an encoded Turing machine $\langle M\rangle$ and a binary word w such that M accepts w. Refer to Lemma 9.3 with regard to encoding Turing machines.*

Lemma 9.4. *The universal language L_u is a recursively enumerable one.*

Proof. We design a Turing machine, which accepts the universal language L_u. This is the so-called universal Turing machine M_u. The machine has three tapes. It realizes the following algorithm

1. checks if an input word on the first tape is of a form $\langle M\ w\rangle$, where $\langle M\rangle$ is a valid code of a Turing machine,
   - looks for beginning sequence of 1s, if it is not of even length $2r$, rejects the input,
   - stores r 0s on the third tape, the content of the third tape is used as a measure of the length of the blocks of binary digits encoding the machine and as a number of the current state,
   - looks for the next sequence of $2r$ digits 1,
   - moves the beginning part of the input word bounded by both blocks of $2r$ digits 1 to the second tape,
   - leaves w on the first tape, places the head of the first tape on the leftmost symbol of w (the second part of input $\langle M\ w\rangle$),
   - checks if the content of the second tape is a valid code $\langle M\rangle$ of a Turing machine,
2. repeats the following actions until the number stored on the third tape encodes the accepting state,
   - for the state q stored on the third tape and for the symbol X stored as a block of r binary digits with the head of the first tape placed on the leftmost digit of this block retrieve a matching transition $\delta(q,X) = (p,Y,D)$. There may be more than one matching transition, so this is nondeterministic choice,
   - replace the content of the third tape with p, replace X by Y and move the head of the first tape to neighboring cell in the direction described by D,
3. accepts the input word once the accepting state is encountered on the third tape. □

Lemma 9.5. *The universal language L_u is not a recursive one.*

Proof. Let us assume that L_u is recursive, i. e., that a Turing machine M' with the stop property accepts L_u. Based on this assumption, the machine M_d accepting the diagonal language in the class *REL* can be built, which contradicts Proposition 9.16. Therefore, the universal language cannot be recursive.

A hypothetical Turing machine M_d, assumed to accept the diagonal language in the class *REL*, realizes the following algorithm:

- for a given input word w, M_d retrieves the index k of w in the canonical order, i. e., it finds such k, that $w = w_k$,
- M_d retrieves binary representation $\langle M_k \rangle$ of the k-th Turing machine M_k,
- M_d concatenates binary representation $\langle M_k \rangle$ of the kth Turing machine M_k with w,
- M_d simulates computation of the hypothetical machine M' for the concatenation of $\langle M_k \rangle$ and w,
- M_d terminates computation if and only if M' terminates its computation, then M_d reverses an output of M', i. e., M_d accepts if and only if M' rejects.

Note that M_d accepts if and only if the Turing machine encoded as k rejects $w = w_k$. Moreover, M_d has the stop property since the hypothetical Turing machine M' is assumed to have the stop property. Therefore, the assumption that M' exists is false. □

Remark 9.6. The following inclusions hold based on discussion in this section

$$RgL \subsetneq CFL \subsetneq CSL \subsetneq RkL \subsetneq REL \subsetneq ALL$$

The following inclusions are called the Chomsky hierarchy

$$RgL \subsetneq CFL \subsetneq CSL \subsetneq REL \subsetneq ALL$$

where $\subsetneq$ denotes inclusion, but not equality.

Part IV: **Advances and applications**

10 Turing machines

10.1 Advances

Problem 10.1. Design a Turing machine accepting prime numbers. Discuss the possibility of building a linear bounded automaton.

Solution. To explain details:
- considered is the set of natural numbers $N = \{0, 1, 2, 3, 4, 5, 6, 7, 8, \ldots\}$, cf. Example 1.1 in Chapter 1,
- 0 and 1 are not prime numbers,
- numbers are represented in the unary system with the digit X.

We design a two-tape Turing machine with the halting accepting and the halting rejecting states, compare Definition 5.11 in Chapter 5. Computation is performed as follows:
1. prologue:
   a. the first tape keeps the input number n and is used for checking if n is a multiple of a number from the second tape,
   b. on the second tape natural numbers $\{2, 3, 4, 5, 6, 7, 8, \ldots, n-1\}$ are consecutively stored,
   c. store the input number n on the first tape, that is store n symbols X, with guards # and & preceding and following it,
   d. check if the input number is 0 or 1, if so, give a correct answer switching to the halting rejecting state (what will halt computation),
   e. copy the content of the first tape to the second tape,
   f. clear the first tape: replace all symbols X on it with the empty tape symbol B,
   g. decrease the number n on the second tape to $n - 1$ (shifting a corresponding guard),
   h. turn content of the second tape to $k = 2$: leave the first two cells with the digit X and clear remaining $n - 3$ cells replacing X with the empty tape symbol B,
2. check if the number n on the first tape is divisible by the number k on the second tape:
   a. copy the content of the second tape (k digits X) to the first tape as many times as possible,
   b. if there is a place for all k digits X in the last copy, then switch to the halting rejecting state: the number n is not a prime one, it is divisible by $1 < k < n$,
3. increase the number k stored on the second tape, clear the first tape, and go to 2.a.,
4. if k stored on the second tape cannot be increased since it reached the value $n - 1$, then switch to the halting accepting state: the number n is prime since it is not divisible by any number $\{2, 3, 4, 5, 6, 7, 8, \ldots, n - 1\}$.

Comment: the constructed Turing machine is a deterministic one.

https://doi.org/10.1515/9783112223123-014

Comment: the machine constructed is not optimal in terms of computational complexity. It is aimed at simplicity of solution. For instance, in point 1.b. consider only numbers $\{4, 5, 6, 7, 8, \ldots, \lfloor n/2 \rfloor\}$ and, after performing point 1.d., additionally check also the input numbers 2 or 3 switching to the halting accepting state.

Comment: notice that on both tapes only their parts of length not greater than the length of the input data are used for computation. Therefore, the machine is a two-tape linear bounded automata. The machine can also be easily transformed to one tape deterministic linear bounded automata: apply methods from Sections 5.1.7 and 5.1.5

Problem 10.2. Design a Turing machine accepting a subset of natural numbers:

$$L = \{k \in N : (\exists p \in N, p \; - \; \text{prime number})\, k + p \text{ is a prime number}\}$$

Does a machine with the stop property exist?

Solution. A solution is based on a programming with Turing machines discussed in Section 5.1.8 of Chapter 5. We will use the Turing machine designed in the previous Problem 10.1 as an element of the solution just to check whether given numbers are prime.

We design a multi-tape Turing machine with the halting accepting state, as defined in Definition 5.11 in Chapter 5. For a given natural number k consecutive prime numbers $p = 2, 3, 5, \ldots$ are identified and checked whether $k + p$ is a prime number:

1. prologue:
   a. numbers are represented in the unary system with the digit X,
   b. the input number k is stored on the first tape,
   c. the first prime number $p = 2$ is stored on the second tape,
2. the third and fourth tapes are used to check if the number p on the second tape is prime using the Turing machine constructed in the above Problem 10.1,
3. if it is not prime then the number p on the second tape is incremented by 1 and computation goes to point 2.,
4. otherwise, if p is prime, the third and fourth tapes are used to check if the sum $k + p$ of numbers on the first and second tapes is prime using again the Turing machine constructed in the Problem 10.1,
5. if the sum $k + p$ is prime, then computation is halted in the accepting state,
6. otherwise the number k on the second tape is incremented by 1 and computation goes to point 2.

Comment: the above Turing machine utilizes a naive method. For a given input number k it iterates over all consecutive natural numbers p, checking which are prime. Then, if a found number p is prime, it checks whether the sum $k + p$ is prime. If such a prime number p exists, then computation halts with a positive answer. On the other hand, if such a prime number does not exist, the computation is infinite. For instance, for $k = 7$, computation is infinite: (i) for the only one even prime number $p = 2$, the sum $k + p = 9$ is not prime and (ii) any prime number p greater than 2 is odd, so then the sum $k + p$ is even, not prime. Therefore, the designed machine does not have the stop property. A question arises whether we can use a non-naive method, an evidence, allowing us to easily determine if, for a given natural number k, such a prime number p exists (a method similar, for instance, to examining the divisibility of k by 11). Unfortunately, we do not know such evidence, but we also do not know if it does not exist.

10.2 Two-dimensional Turing machines

There are numerous varieties of Turing machines, many of which were discussed in Chapter 5, for instance, machines with multi-track tapes and multi-tape Turing machines. Classes of different variants are equivalent to each other in terms of languages accepted or problems solved.

In this section, we discuss another generalization, that is, two-dimensional Turing machines or, in other words, Turing machines with two-dimensional tape. We also justify that the class of two-dimensional Turing machines is equivalent to one-dimensional ones (as studied in Chapter 5, different varieties of one-dimensional ones are equivalent to each other). Following this path, it would be easy to study and prove the equivalence of multi-dimensional Turing machines.

Definition 10.1. A description of a two-dimensional Turing machine would be easily adapted from Definition 5.1.1:

$$M = (Q, \Sigma, \Gamma, \delta, q_0, B, F, C)$$

The difference is in the definition of the transition function, which must take into account transitions of a head in four directions:

$$\delta : Q \times \Gamma \to Q \times \Gamma \times \{L, R, D, U\}$$

where L, R, D, U denote left, right, down and up directions.

Alike in the case of a one-dimensional model, the transition function δ may not be a total function, that is, it may be undefined for some of its arguments. Such a case is formally interpreted as the machine falls into an infinite computation.

A two-dimensional Turing machine could be interpreted as a physical mechanism shown in Figure 10.1. This mechanism consists of:

- a control unit, it is in a state of Q,
- a four-way infinite tape split into cells; every cell contains a symbol (exactly one) of the tape alphabet Γ,
- the head, it is placed over a cell of a tape,

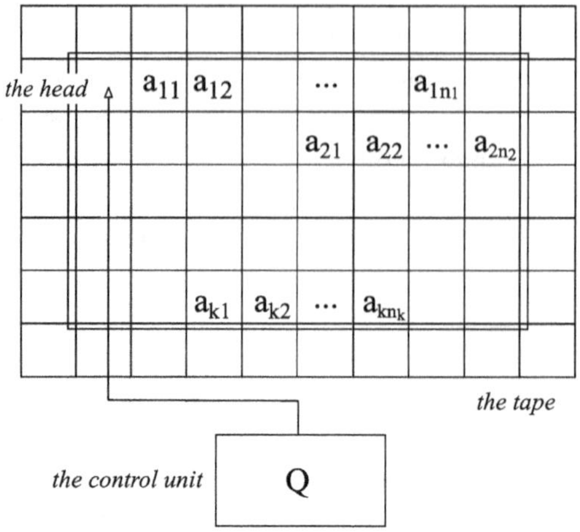

Figure 10.1: Turing machine with two-dimensional tape.

- to make a move, the machine checks a state of the control unit and the head reads a symbol held in the cell under it. Then the machine applies the transition function as explained below.

Turing machines perform computations for a given input data. The computation of a given Turing machine is done according to the following intuitive procedure:

1. the initial configuration of a given Turing machine is described as follows:
   a. input data, a word $w = a_1 a_2 \dots a_n$ over input alphabet Σ, is stored in n consecutive cells of a row of the tape,
   b. all other cells of the tape (which is infinite to four directions) are filled in with the blank symbol B,
   c. the head of the Turing machine is placed over the cell with the first (leftmost) symbol a_1 of the input word w,
   d. the control unit is in the beginning state q_0,
2. if the stop condition C is satisfied, then the computation is halted, the configuration is called the final configuration, the machine responses whether its control unit is in an accepting state or not,
3. if the stop condition C is not satisfied, then – based on the state q of the control unit and the symbol X read by the head – the Turing machine is doing the following actions:
   a. the value (p, Y, D) of the transition function $\delta(q, X)$ is computed,
   b. the head stores the tape symbol Y in the cell under it,
   c. the control unit switches to the state p,
   d. the head shifts by one cell in the direction D,
4. computation proceeds to point 2.

10.3 Two-dimensional versus one-dimensional Turing machines

Two-dimensional Turing machines are equivalent to one-dimensional ones in the sense that computations of a machine of one type can be simulated by a machine of another type. In other words, any problem that could be solved by a machine of one type can also be solved by a machine of another type.

Obviously, a one-dimensional machine is a special case of a two-dimensional one. An equivalent two-dimensional one performs computation just on one row according to a transition function with values limited to only left and right shifts.

Talking about the simulation of a two-dimensional machine by a one-dimensional one, notice that at any stage of computation, only a finite part of the tape is filled in with a non-blank symbol. The number of non-blank cells, as well as the number of cells visited by the head, is limited by the length of the input data plus the number of moves of the machine. This observation concerns any type of Turing machine. Discussing two-dimensional Turing machines, only the minimal rectangle will be considered that includes all cells with input data stored in and all cells visited by the head, cf. Figures 10.1 and 10.2.

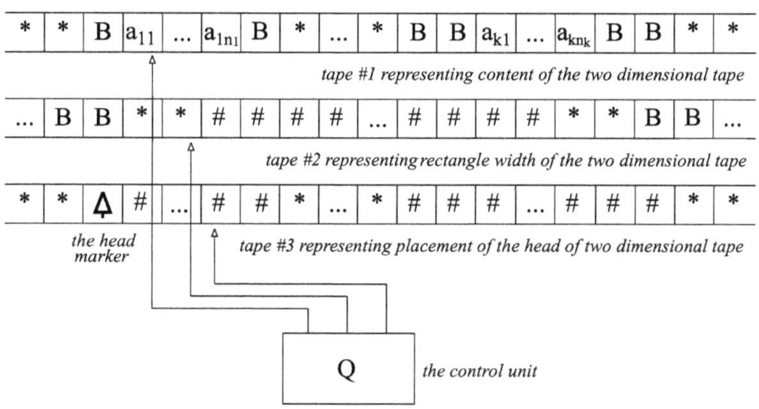

Figure 10.2: Representation of a two-dimensional Turing machine exposed in Figure 10.1 by a one-dimensional Turing machine with three tapes.

We will represent a two-dimensional tape using three single-track tapes of one dimension, unbounded on both sides:

- the state of the two-dimensional tape will be recorded on a one-dimensional tape by writing successive rows of the rectangle described above. The rows will be separated by a special symbol (denoted by *) not belonging to the tape alphabet of the two-dimensional tape machine. The beginning and end of the recording on the one-dimensional tape will be marked with a double special symbol, see Figure 10.2,
- the second tape of the simulating machine is used to record the width (the length of a line) of the two-dimensional tape rectangle,

- the third tape is used to record the head position of the first tape, and it is required when the width of the rectangle on the two-dimensional tape is increased.

Simulation of a move of a two-dimensional machine is based on the rectangle described above and is performed as follows:
- move of the head to the left or right, if it does not go out of the rectangle, is simulated by the move of heads of the first and third tapes to the left or right,
- move of the head to the left or right with the head going out of the rectangle requires:
  - adding a new cell to each fragment of the first tape corresponding to rows of the rectangle and going back with the head to the initial position (stored in the third tape),
  - adding a new cell to each fragment of the third tape corresponding to rows of the rectangle and going back with the head to the initial position (stored in the first tape),
  - adding a new cell to the fragment of the second tape representing the width of the rectangle,
  - making a move as outlined in the above point,
- move of the head up or down, if it does not go outside the rectangle as mentioned above, corresponds to the move of the head of the first tape left or right by the length of the rectangle width (plus one, taking into account the separator). The action of moving the head uses the information about the rectangle width recorded on the second tape,
- if the head moves up or down outside the rectangle, it means that the height of the rectangle is increased. It requires adding a record to the first and third tapes, a record representing a new row of the rectangle, and then making a move as described in the point above,
- of course, the actions from the previous points end with switching to the state and printing an alphabet symbol corresponding to a state and a symbol of a move of the two-dimensional machine.

10.4 The hierarchy revisited

In Section 9.3 we outlined the hierarchy of languages based on inclusions of them. It is the so-called Chomsky hierarchy supplemented with the class of recursive languages. This section gives another view on the class of all languages. As outlined in Figure 10.3, we distinguish the following classes:

RgL regular languages,
CFL context-free languages,
CSL context-sensitive languages,
RkL recursive languages,
REL recursively enumerable languages,

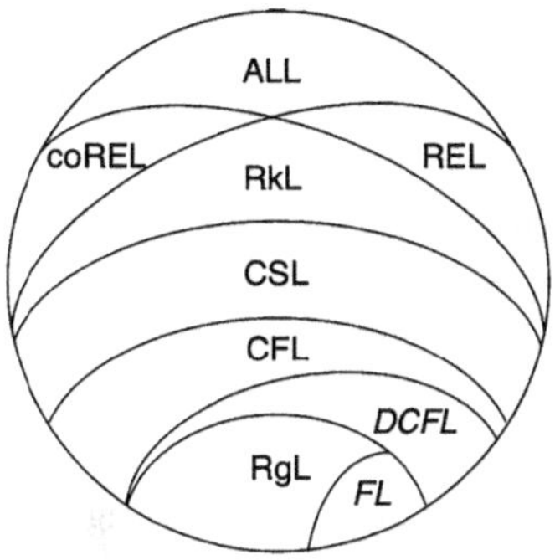

Figure 10.3: The hierarchy of languages.

coREL languages complimentary to recursively enumerable ones, and
ALL all languages.

Comparing to the hierarchy discussed in Chapter 9, we have one new class – languages complementary to recursively enumerable, denoted *coREL*. This language is discussed in more detail further below.

Let us shortly explain the class *FL*, which includes only finite languages. Finite languages are regular ones and this class is easily identified. Justification of this note is trivial: any finite language is generated by a regular expression, being the sum of all its words.

The class *DCFL* includes context-free languages that are accepted by deterministic push-down automata. We have context-free languages that are accepted by deterministic push-down automata and that are accepted by push-down automata but not by deterministic ones. Unfortunately, we do not know a rule identifying this class.

Since the set of Turing machines is countable, cf. Lemma 9.3, so that the class of recursively enumerable languages, *REL*, is also countable (because they correspond to Turing machines). Therefore, the class of languages complementary to recursively enumerable, *coREL*, is also countable. And, as we see, the class of recursive languages, *RkL*, is the intersection of both classes *REL* and *coREL*. In other words, if a language is recursively enumerable and its complement is also recursively enumerable, then both are recursive.

In Chapter 9, we outlined the universal language, L_u, as recursively enumerable, but not recursive. From the discussion in that Chapter, we conclude that the complementary language to the universal one, $\overline{L_u}$, is not recursively enumerable. We also proved that the diagonal language, L_d, is not recursively enumerable. On the other hand, the complementary language to the diagonal one, $\overline{L_d}$, is recursively enumerable.

Therefore, we can initially populate classes *REL* and *coREL* as follows:
- L_u, $\overline{L_d} \in REL$,
- $\overline{L_u}$, $L_d \in coREL$.

The class *ALL* includes more languages than in both classes *REL* and *coREL*. Let us recall that this is a direct consequence of the facts that classes *REL* and *coREL* are countable and the class ALL is uncountable. Now, we present two languages that are neither recursively enumerable, nor complementary to them, i. e., we populate the class above REL and $coREL$, cf. Figure 10.3.

Consider the set of all words over the binary alphabet $\{0,1\}^*$. We interpret those words as codes of Turing machines, cf. Lemma 9.3 in Chapter 9. It is straightforward to check if a binary word is a correct code of a Turing machine, and if not, we can assume that such a word encodes a Turing machine doing infinite computation for any of its input. Notice that the machine accepts the empty language (which is a recursive one).

Now, let us distinguish the language L_r of binary words being codes of Turing machines accepting recursive languages. All other words create the language L_{nr} complementary to L_r. This language includes codes of Turing machines that accept languages that are not recursive. These two languages, i. e.:
- codes of Turing machines accepting recursive languages, and
- codes of Turing machines accepting non-recursive languages

are neither in the class REL, nor in $coREL$.

To prove it, let us design two auxiliary Turing machines $M'(M,w)$ and $M''(M,x)$ that depend on *two parameters*: a Turing machine M and a word w. They also *include* a universal Turing machine M_u that accepts the universal language L_u (it is defined in 9.7). These machines are displayed in Figure 10.4.

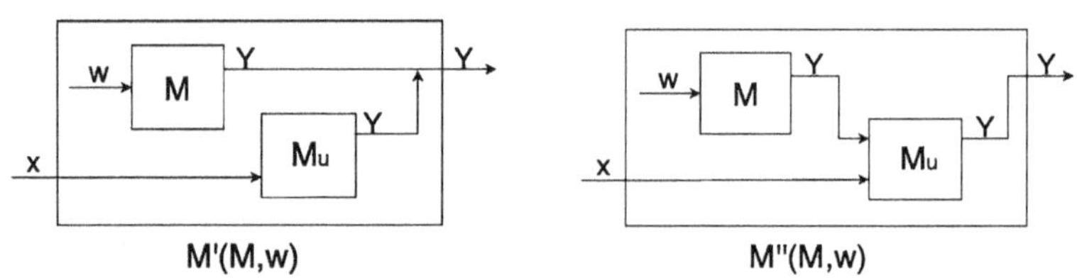

Figure 10.4: Auxiliary Turing machines $M'(M,w)$ and $M''(M,w)$ that depend on two parameters: a Turing machine M and a word w. M_u is a universal Turing machine.

The machine $M'(M,w)$ runs computations of both *internal* machines for their inputs: M for w and M_u for x. M' halts computation and accepts its input x if and only if M or M_u halts computation and accepts associated input. The machine accepts either the set of all word Σ^* or the universal language L_u:

$$L(M') = \begin{cases} \Sigma^* & \text{if } M \text{ accepts } w \\ L_u & \text{otherwise} \end{cases}$$

The machine $M''(M,w)$, analogously to the machine $M'(M,w)$, runs computation of its *internal* machine M for w and then, if M halts computation and accepts its input w, it starts computation of M_u for x. M'' halts computation and accepts its input x if and only if M_u halts and accepts associated input x. Therefore the machine accepts either the universal language L_u or the empty language $\emptyset$:

$$L(M') = \begin{cases} L_u & \text{if } M \text{ accepts } w \\ \emptyset & \text{otherwise} \end{cases}$$

Now, let us suppose that languages L_{nr} and L_r are from $REL \cup coREL$, i. e., at least one of them is recursively enumerable, i. e., there is at least one Turing machine out of M_{nr} and M_r corresponding to these languages. If this assumption is true, then it would be possible to construct a Turing machine accepting the diagonal language L_d. Such a construction is displayed in Figure 10.5.

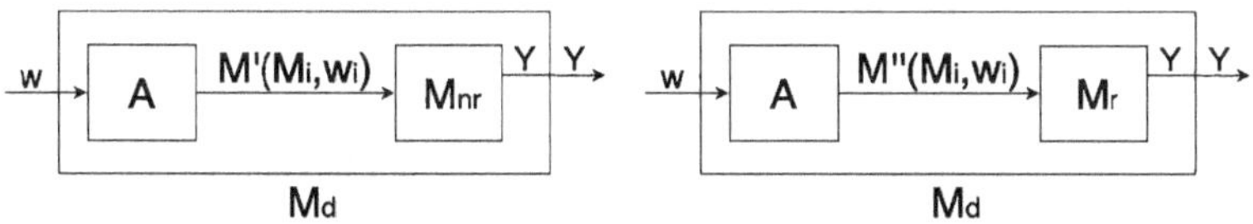

Figure 10.5: A hypothetical Turing machines M_d accepting the diagonal language L_d in the class of recursively enumerable languages assuming that machines M_{nr} and M_r exist.

The hypothetical left machine M_d outlined in Figure 10.5 works as follows:
- it takes an input w and presents it to an internal algorithms (a Turing machine with stop property) A,
- A retrieves index n of w in the canonical order of words and then constructs a Turing machine $M'(M_n, w)$,
- M_d presents code of $M'(M_n, w)$ to a machine M_{nr} and propagates its output.

Since $M'(M, w)$ accepts the universal language when $w \notin M$, then $M'(M_n, w)$ accepts any word $w \equiv w_n$ such that M_n does not accept it, then M_d would accept the diagonal language L_d, which is not possible. Therefore, L_{nr} is not recursively enumerable.

We can show that L_r is not recursively enumerable, analogously referring to the right construction in Figure 10.5.

11 Context-free grammars – a real application

11.1 Applications

This section is devoted to selected applications of context-free grammars. The section is a roadside of the main discussion on formal languages, automata and computability. Topics included in this section are a small part of a compiler's practice. They can be used in an elementary project on parsing basic constructions like arithmetic expressions, which are among the most complex parts of programming languages. This section is not aimed at a complete and detailed presentation of parsing. It is rather a signalization of the theme.

11.1.1 Translation grammars

This section is focused on some modifications of context-free grammars. Despite that modifications presented here are not included in the main flow of discussion on context-free grammars, the associated practical importance makes them valuable and justifies their presentation in the book. Two extensions of context-free grammars are presented, namely translation grammars and LL(1) grammars. These types of grammars will be used to parse arithmetic expressions and translate them to postfix form.

Translation grammars stem from context-free grammars. They could be seen as context-free grammars with the simultaneous derivation of related words.

Definition 11.1. Translation grammar is a context-free grammar $G = (V, T, P, S)$ with the set T of terminal symbols split into two disjoint subsets T' and T'' of primary and secondary symbols, i. e., $T = T' \cup T''$, $T' \cap T'' = \emptyset$.

A translation grammar is a context-free grammar producing a context-free language $L(G)$ over the alphabet T of terminal symbols. On the other hand, we can say that the translation grammar produces two languages: the primary language $L'(G)$ and the translation language $L''(G)$. The primary language is obtained from the language $L(G)$ by removing translation symbols from its words. Then again, translation language is obtained from the language $L(G)$ by removing primary symbols from its words.

Example 11.1. The following grammar generates nonempty words over the alphabet $\Sigma = \{a, b\}$ having the same numbers of symbols a and b, cf. Problem 3.1.

$$G = (\{S, A, B\}, \{a, b\}, \{S \to aB \mid bA, A \to a \mid aS \mid bAA, B \to b \mid bS \mid aBB\}, S)$$

Solution. The following translation grammar G_T generates the primary language $L'(G)$ equal to $L(G)$ and the translation language $L''(G)$, which has words with symbols a shifted to the left:

$$G_T = (\{S, A, B\}, \{a, b, \overline{a}, \overline{b}\})$$

https://doi.org/10.1515/9783112223123-015

where we have the following set $P(T)$ of productions:

$$
\begin{aligned}
P_T\colon\quad & S \to \overline{a}aB \mid bA\overline{b} \\
& A \to \overline{a}a \mid \overline{a}aS \mid bAA\overline{b} \\
& B \to b\overline{b} \mid bS\overline{b} \mid \overline{a}aBB
\end{aligned}
$$

The set of terminal symbols $\{a, b, \overline{a}, \overline{b}\}$ has two primary symbols $\{a, b\}$ and the set of translation symbols $\{\overline{a}, \overline{b}\}$. Productions of the grammar G_T directly correspond to the productions of the grammar G.

Let us consider the word $w = baba$. Its derivation tree (in the grammar G) is shown in the left part of Figure 11.1. The corresponding derivation tree in the grammar G_T is shown in the right part of Figure 11.1. Its crop is equal to $w_T = b\overline{a}ab\overline{a}a\overline{b}\overline{b}$ and is split to the primary word $w' = baba$ and the secondary word $w'' = aabb$.

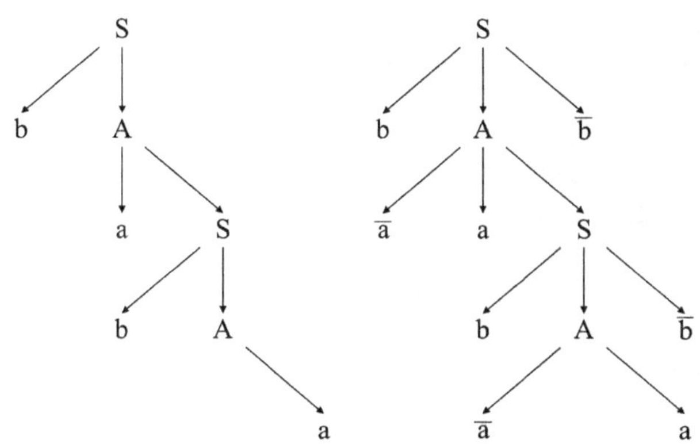

Figure 11.1: The derivation trees obtained in a translation grammar.

This example can be interpreted as follows. For the given grammar G and a word w, we can find its derivation in the grammar G. Then we design the corresponding derivation in the grammar G_T. The generated word w_T is split to the primary w', which is equal to the word w, and the secondary w'', which is the translation of its primary counterpart w'.

11.1.2 LL(1) grammars

In this section, we assume that, for a context-free grammar $G = (V, T, P, S)$, any word $w \in T^*$ has a special end-of-word symbol $\lhd$ appended. This symbol is neither a nonterminal symbol nor a terminal symbol. It is used for marking the end of any intermediate and terminal word of a derivation of the word w.

Assume that $p : A \to \alpha$ is a production in a context-free grammar $G = (V, T, P, S)$, where $A \in V, \alpha \in \{V \cup T\}^*$. Let us define the following sets of symbols:

- $FIRST(p) = FIRST(\alpha)$ is the set of those terminal symbols, which may open any intermediate word derivable from α. Note that if the right hand side α of the production p begins with a terminal symbol a, then $FIRST(p) = FIRST(\alpha) = \{a\}$. On the other hand, $FIRST(1) = FIRST(L) = \{a\}$ for the production $1 : S \to L$ of the grammar of Example 3.3,
- $FOLLOW(p) = FOLLOW(A)$ is the set of those terminal symbols and end-of-word symbol, which may directly follow A in any intermediate word derived from the beginning symbols S of the grammar G. Note that $FOLLOW(3) = FOLLOW(L) = \{c, \lhd\}$ for the production $3 : L \to Lc$ of the grammar of Example 3.3,
-

$$SELECT(p) = \begin{cases} FIRST(p) \cup FOLLOW(p) & \text{if } p \text{ is nullable} \\ FIRST(p) & \text{otherwise} \end{cases}$$

Definition 11.2. A context-free grammar $G = (V, T, P, S)$ is $LL(1)$ grammar if and only if for every nonterminal symbol $A \in V$ all A-productions have SELECT sets pairwise disjoint. This condition is called LL(1) uniqueness condition.

LL(1) grammars are tools for building a top-down membership analyzer (top-down parser). LL(1) grammars are tools for designing **L**eftmost derivation of the input word, processing the word from **L**eft to right. The derivation is designed based on **1** input symbol at a time.

Note that the uniqueness condition of the LL(1) grammars is similar to the Greibach uniqueness condition. This allows for an easy designing of the derivation of a word: for a given leftmost nonterminal symbol $A \in V$ in an intermediate derivation word and a given input symbol $a \in T$ we apply this A-production to the nonterminal A, which has the terminal a in its SELECT set. When the right-hand side of the applied production begins with the terminal symbol a, the input is shifted to the next input symbol. Translation symbols in intermediate derivation words are skipped during this processing.

Example 11.2. Let us consider the following grammar generating arithmetic expressions with addition, subtraction, multiplication, division, power, change of sign (negation) and brackets:

$$G = (\{E, E_l, T, T_l, P, P_l, Q\}, \{+, -, *, /, \uparrow, (,), id\}, P, E)$$

The set of productions is given in Table 11.1. Productions are supplemented with FIRST, FOLLOW and SELECT sets (FOLLOW sets are needed only for nullable productions). SELECT sets show that the grammar is LL(1) one.

Example 11.3. Below we present a translation grammar, which converts arithmetic expressions to a postfix form (Reverse Polish Notation). The translation grammar stems from the grammar G of Example 11.2. The set of terminal symbols of the grammar is supplemented with translation symbols $T'' = \{\overline{+}, \overline{-}, \equiv, \overline{*}, \overline{/}, \overline{\uparrow}, \overline{id}\}$. Productions are supplemented with translation symbols as shown in Table 11.2. The uniqueness property of the

Table 11.1: A LL(1) grammar generating arithmetic expressions.

No.	Production	FIRST	FOLLOW	SELECT
1.	$E \to TE_I$	$\{-, (, id\}$		$\{-, (, id\}$
2.	$E_I \to +TE_I$	$\{+\}$		$\{+\}$
3.	$E_I \to -TE_I$	$\{-\}$		$\{-\}$
4.	$E_I \to \varepsilon$	$\emptyset$	$\{), \lhd\}$	$\{), \lhd\}$
5.	$T \to PT_I$	$\{-, (, id\}$		$\{-, (, id\}$
6.	$T_I \to *PT_I$	$\{*\}$		$\{*\}$
7.	$T_I \to /PT_I$	$\{/\}$		$\{/\}$
8.	$T_I \to \varepsilon$	$\emptyset$	$\{+, -,), \lhd\}$	$\{+, -,), \lhd\}$
9.	$P \to QP_I$	$\{-, (, id\}$		$\{-, (, id\}$
10.	$P_I \to \uparrow QP_I$	$\uparrow$		$\{\uparrow\}$
11.	$P_I \to \varepsilon$	$\emptyset$	$\{+, -, *, /,), \lhd\}$	$\{+, -, *, /,), \lhd\}$
12.	$Q \to -Q$	$\{-\}$		$\{-\}$
13.	$Q \to (Q)$	$\{(\}$		$\{(\}$
14.	$Q \to id$	$\{id\}$		$\{id\}$

Table 11.2: A LL(1) grammar generating arithmetic expressions and their conversion to postfix form.

No.	Production	FIRST	FOLLOW	SELECT
1.	$E \to TE_I$	$\{-, (, id\}$		$\{-, (, id\}$
2.	$E_I \to +T \overline{+} E_I$	$\{+\}$		$\{+\}$
3.	$E_I \to -T \overline{-} E_I$	$\{-\}$		$\{-\}$
4.	$E_I \to \varepsilon$	$\emptyset$	$\{), \lhd\}$	$\{), \lhd\}$
5.	$T \to PT_I$	$\{-, (, id\}$		$\{-, (, id\}$
6.	$T_I \to *P \overline{*} T_I$	$\{*\}$		$\{*\}$
7.	$T_I \to /P \overline{/} T_I$	$\{/\}$		$\{/\}$
8.	$T_I \to \varepsilon$	$\emptyset$	$\{+, -,), \lhd\}$	$\{+, -,), \lhd\}$
9.	$P \to QP_I$	$\{-, (, id\}$		$\{-, (, id\}$
10.	$P_I \to \uparrow QP_I \overline{\uparrow}$	$\uparrow$		$\{\uparrow\}$
11.	$P_I \to \varepsilon$	$\emptyset$	$\{+, -, *, /,), \lhd\}$	$\{+, -, *, /,), \lhd\}$
12.	$Q \to -Q \equiv$	$\{-\}$		$\{-\}$
13.	$Q \to (Q)$	$\{(\}$		$\{(\}$
14.	$Q \to id\ \overline{id}$	$\{id\}$		$\{id\}$

LL(1) grammar is inherited from the property of the original grammar of Example 11.2. Since this property is associated with the primary language, so then FIRST, FOLLOW and SELECT sets are unaffected and stay the same as in the grammar of Example 11.2.

The set of productions can be split into several groups concerning their functions:

- productions 1–4 generate additive operations (addition, subtraction),
- productions 5–8 generate multiplicative operations (multiplication, division),
- productions 9–11 generate power operation,
- production 12 generates change of sign operation,

- production 13 generates brackets,
- production 14 generates arguments of operations (numbers, variables) represented by the terminal symbol *id*.

This split of operations reflects the precedence of operators. The following hierarchy is utilized when an expression is evaluated:
- additive operators (addition and subtraction) have the lowest priority,
- multiplicative operators (multiplication and division) have higher priority than additive operators do,
- the power operator has higher priority than multiplicative operators do,
- the change of sign operators has the higher priority than the power operator does,
- brackets give the highest priority for included sub-expression,
- operators of the same priority are evaluated:
  - left to right in case of additive operators,
  - left to right in case of multiplicative operators,
  - right to left in case the power and the change of sign operator.

This hierarchy is preserved in the analysis of arithmetic expressions and their conversion to other forms.

Analysis of next three arithmetic expressions employs the grammar with productions outlined in Table 11.2 with production number 14 replaced with the following ones.

No.	Production	FIRST	FOLLOW	SELECT
14.	$Q \to a\,\overline{a}$	$\{a\}$		$\{a\}$
	$Q \to b\,\overline{b}$	$\{b\}$		$\{b\}$
	$Q \to c\,\overline{c}$	$\{c\}$		$\{c\}$
	$Q \to d\,\overline{d}$	$\{d\}$		$\{d\}$

The derivation of the expression $-a + b * c - d$ is designed as follows:
- we start from the initial symbol E of the grammar of Example 11.3,
- there is only one E-production to be applied to the input symbol – and the initial symbol of the grammar E:

$$E \xrightarrow{1} TE_l,$$

- there is only one T-production to be applied to the input symbol – and the leftmost nonterminal T:

$$E \xrightarrow{1} TE_l \xrightarrow{5} PT_lE_l,$$

- there is only one P-production to be applied to the input – and to the leftmost nonterminal P:

$$E \xrightarrow{1} TE_l \xrightarrow{5} PT_lE_l \xrightarrow{9} QP_lT_lE_l,$$

- there are three Q-productions to be applied to the input symbol – and the leftmost nonterminal Q, but SELECT set of only one Q-production includes the input symbol – and this production is applied:

$$E \xrightarrow{*} QP_lT_lE_l \xrightarrow{12} -Q \equiv P_lT_lE_l$$

- since in the previous step the input symbol – has been added to the intermediate derivation word by the production number 12, now input is shifted to the symbol a and there are three Q-productions to apply to the input symbol and the leftmost nonterminal Q, but SELECT set of only one Q-production includes the input symbol a and this production is applied:

$$E \xrightarrow{*} -Q \equiv P_lT_lE_l \xrightarrow{14} -a \equiv P_lT_lE_l,$$

- now input is shifted to the symbol +, the translation symbol ≡ in the intermediate word of derivation is skipped, so then there are three P_l-productions to apply to the input symbol + and the leftmost nonterminal P_l, but SELECT set of only one P_l-production includes the input symbol + and this production is applied:

$$E \xrightarrow{*} -a \equiv P_lT_lE_l \xrightarrow{11} -a \equiv T_lE_l,$$

- there are three T_l-productions to apply to the input symbol + and the leftmost nonterminal T_l, but SELECT set of only one T_l-production includes the input symbol + and this production is applied:

$$E \xrightarrow{*} -a \equiv P_lT_lE_l \xrightarrow{11} -a \equiv T_lE_l \xrightarrow{8} -a \equiv E_l,$$

- there are three E_l-productions to apply to the input symbol + and the leftmost nonterminal E_l, but SELECT set of only one E_l-production includes the input symbol + and this production is applied:

$$E \xrightarrow{*} -a \equiv T_lE_l \xrightarrow{8} -a \equiv E_l \xrightarrow{2} -a \equiv \ +T \mp E_l,$$

- now input is shifted to the symbol b and there is only one T-productions to apply to the input symbol and the leftmost nonterminal T:

$$E \xrightarrow{*} -a \equiv E_l \xrightarrow{2} -a \equiv \ +T \mp E_l \xrightarrow{5} -a \equiv +PT_l \mp E_l,$$

- etc.

The derivation tree equivalent for the above derivation is shown in Figure 11.2. The primary word of this derivation tree is – of course – $w' = w = -a + b * c - d$. The secondary (translated) word is equal to $w'' = \overline{a} \mathbin{\overline{=}} \overline{b}\ \overline{c}\ \overline{*}\ \overline{+}\ \overline{d}\ \overline{-}$. Note that ordinary symbols are marked with an upper dash in secondary word. Moreover, subtract operator and change of sign operator are distinguished in the secondary word. The former one is shown in the form of an overlined minus sign $\overline{-}$ and the latter is shown as an overlined equality sign $\overline{=}$. A distinction between these two operators is required since their meaning cannot be drawn from the context of expression in postfix form. Unlike, expressions in traditional form (viz. infix form) allow for differentiating meaning of the minus sign, which stands for both operators: subtraction and changing of the sign. Two additive operators are utilized in order from the left one to the right one, i. e., they are left-to-right associative.

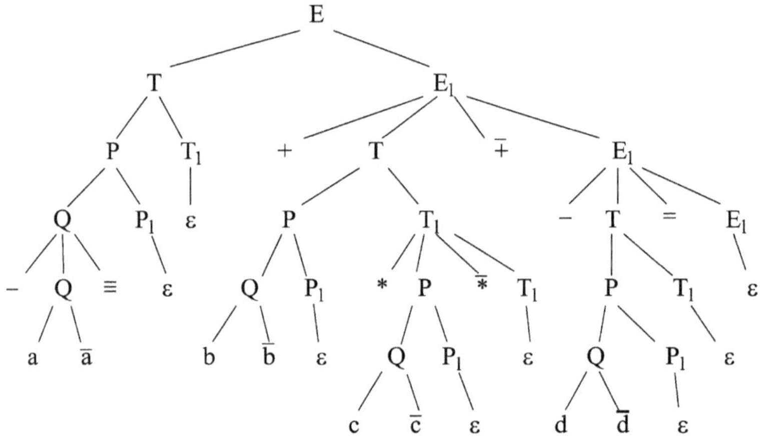

Figure 11.2: The derivation tree of the expression $-a + b * c - d$ in the LL(1) translation grammar of Example 11.3 and the translated expression $\overline{a} \mathbin{\overline{=}} \overline{b}\ \overline{c}\ \overline{*}\ \overline{+}\ \overline{d}\ \overline{-}$.

Unlike in the infix form, arithmetic expressions in postfix form do not need brackets to change priorities of operators. This property is illustrated in Figure 11.3. The primary word of this derivation tree is $w' = w = -(a + b) * c - d$. The secondary (translated) word is equal to $w'' = \overline{a}\ \overline{b}\ \overline{+} \mathbin{\overline{=}} \overline{c}\ \overline{*}\ \overline{d}\ \overline{-}$.

In Figure 11.4 the right to left associativity of power and change of sign operators is illustrated. The primary word of this derivation tree is $w' = w = - - a \uparrow - - b \uparrow c$. The secondary (translated) word is equal to $w'' = \overline{a} \mathbin{\overline{=}} \mathbin{\overline{=}} \overline{b} \mathbin{\overline{=}} \mathbin{\overline{=}} \overline{c}\ \overline{\uparrow}\,\overline{\uparrow}$.

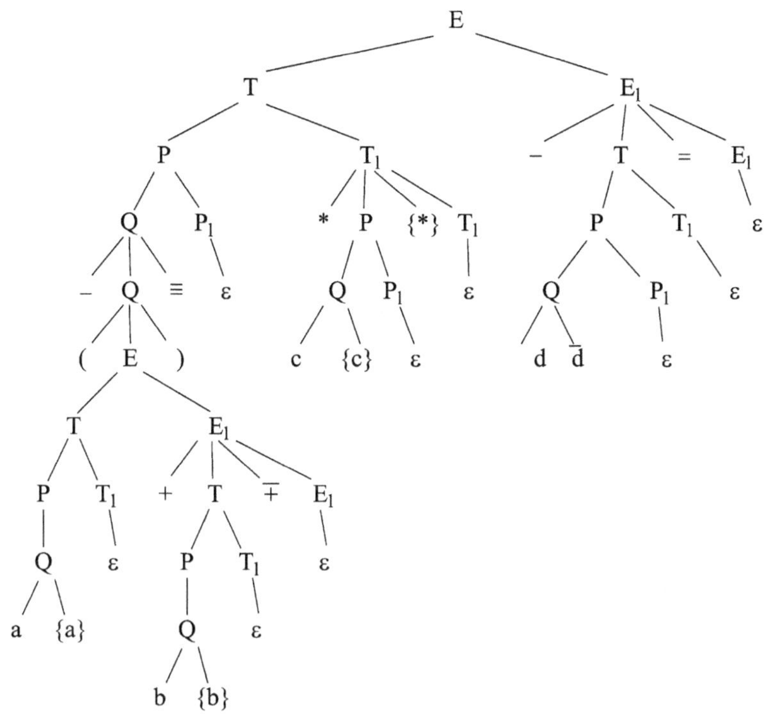

Figure 11.3: The derivation tree of the expression $-(a + b) * c - d$ in the LL(1) translation grammar of Example 11.3 and the translated expression $\overline{a}\,\overline{b}\,\overline{+} \equiv \overline{c}\,\overline{*}\,\overline{d}\,\overline{-}$.

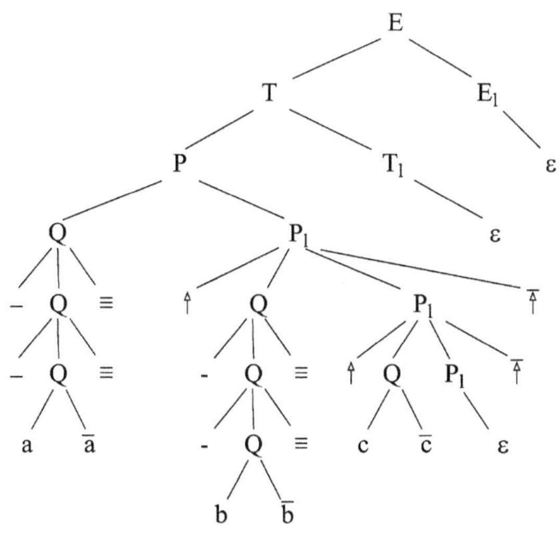

Figure 11.4: The derivation tree of the expression $- - a \uparrow - - b \uparrow c$ in the LL(1) translation grammar of Example 11.3 and the translated expression $\overline{a} \equiv \equiv \overline{b} \equiv \equiv \overline{c}\,\overline{\uparrow}\,\overline{\uparrow}$.

12 Finite automata

12.1 Applications

In this chapter, we illustrate the detection of irregularities and abnormalities using informative and straightforward examples, based on deterministic and nondeterministic classical automata, as described in [5]. Then we generalize the description of classical automata to include automata with imperfect information, cf. [3]. This generalized description can be applied to different types of automata, for instance, fuzzy finite automata, cf. [1, 2, 6, 8]. Finally, we illustrate the accumulation of irregularities using fuzzy automata with arbitrarily chosen t-norms and t-norms viewed as information aggregators. The terms "aggregation" and "accumulation" are used interchangeably. A slight, however significant difference, remains in the sense that accumulation is a kind of strictly monotonic aggregation.

The discussion is structured as follows. In Section 12.2, notions basic for discussion on the main topic of this study are referred to: fuzzy sets with generalized fuzzy connectives, aggregators of information, and finite automata with imperfect information. In Section 12.3, we illustrate step by step the process leading to the accumulation of irregular input data. Finally, in Section 12.4 we summarize the discussion and outline future research directions.

12.2 Preliminaries

In this section, we cover some prerequisites concerning fuzzy sets including the extension of fuzzy operators to triangular norms, aggregating operators, and finite automata, as well as their generalizations to process imperfect information.

12.2.1 Fuzzy sets and triangular norms

Basic notions of fuzzy sets and generalization of fuzzy connectives, maximum and minimum, to triangular norms are considered. We express fuzzy sets in the form of membership functions. A fuzzy set A defined in the universe X is a mapping $\mu : X \to [0,1]$ or $\mu_{A,X} : X \to [0,1]$ if the semantics of the set and the universe are explicitly stated.

The Zadeh's model of fuzzy sets can be described as a system similar to set theory $(F(X), \cup, \cap, ^-)$, where $F(X)$ denotes fuzzy sets over the universe X and $\cup$, $\cap$, $^-$ denote union, intersection, and complement. This system is clearly interpreted as $([0,1]^X, \max, \min, 1-)$, where $[0,1]^X$ denotes all mappings from the universe X into the unit interval $[0,1]$, i. e., the space of membership functions, and max, min and (1−) applied to membership functions implement union, intersection and complement. We ignore the interpretation of fuzzy sets in terms of a lattice L^X.

https://doi.org/10.1515/9783112223123-016

Triangular norms are a useful generalization of basic operators on fuzzy sets realized by the max and min operators.

Let us recall that triangular norms (t-norms and t-conorms) are mappings from the unit square $[0,1] \times [0,1]$ onto the unit interval $[0,1]$ satisfying axioms of associativity, commutativity, monotonicity, and boundary conditions (cf. [7] for details), i. e.:

t-norms and t-conorms are mappings $p : [0,1] \times [0,1] \rightarrow [0,1]$, where p stands for both t-norm and t-conorm, satisfying the following axioms:

1. $p(a, p(b,c)) = p(p(a,b),c)$ – associativity
2. $p(a,b) = p(b,a)$ – commutativity
3. $p(a,b) \leq p(c,d)$ if $a \leq c$ and $b \leq d$ – monotonicity
4. $t(1,a) = a$ for $a \in [0,1]$ – t-norm boundary condition
   $s(0,a) = a$ for $a \in [0,1]$ – t-conorm boundary condition

t-norms and t-conorms are dual operations in the sense that for any given t-norm t, we have a dual t-conorm s defined by the De Morgan formula:

$$s(a,b) = 1 - t(1-a, 1-b)$$

And vice-versa, for any given t-conorm s, we have a dual t-norm t defined by the De Morgan law

$$t(a,b) = 1 - s(1-a, 1-b)$$

The duality of triangular norms causes the duality of their properties. Note that the min/max is a pair of dual t-norm and t-conorm. Some selected known t-norms and their dual t-conorms are given in Table 12.1. Note that the Hamacher product is the chosen representative of the parametric class of Hamacher norms.

Table 12.1: Selected triangular norms, dual t-norms and t-conorms are placed in one row.

t-norm		t-conorm	
minimum	$\min(x,y)$	maximum	$\max(x,y)$
product	$x \cdot y$	probabilistic sum	$x + y - x \cdot y$
Lukasiewicz	$\max(0, x+y-1)$	bounded sum	$\min(a+b, 1)$
nilpotent minimum	$\begin{cases} \min(x,y) & \text{if } x+y>1 \\ 0 & \text{otherwise} \end{cases}$	nilpotent maximum	$\begin{cases} \max(x,y) & \text{if } x+y<1 \\ 1 & \text{otherwise} \end{cases}$
drastic	$\begin{cases} y & \text{if } x=1 \\ x & \text{if } y=1 \\ 0 & \text{otherwise} \end{cases}$	drastic	$\begin{cases} y & \text{if } x=0 \\ x & \text{if } y=0 \\ 1 & \text{otherwise} \end{cases}$
Hamacher product	$\begin{cases} 0 & \text{if } x=0=y \\ \frac{x \cdot y}{x+y-x \cdot y} & \text{otherwise} \end{cases}$	Einstein sum	$\frac{x+y}{1+x \cdot y}$

t-norms and t-conorms are bounded by the minimum t-norm and the maximum t-conorm, that is, for any t-norm t, any t-conorm s, and any $x, y \in [0,1]$:

$$t(x,y) \leq \min(x,y) \leq \max(x,y) \leq s(x,y) \tag{12.1}$$

12.2.2 Aggregation operators

Finite automata are used to aggregate information. For instance, some irregularities appearing in input data may be accumulated in order to control the level of such irregularities. Here the interpretation of aggregation of information given in [7] is recalled.

An *aggregating operator* is a mapping $\mathcal{A} : \bigcup_{n \in \mathcal{N}} [0,1]^n \to [0,1]$ such that:

1. $\mathcal{A}(x_1, \ldots, x_n) \leq \mathcal{A}(y_1, \ldots, y_n)$ if $x_i \leq y_i$ for $i = 1, \ldots, n$,
2. $\mathcal{A}(x) = x$ for all $x \in [0,1]$,
3. $\mathcal{A}(0, \ldots, 0) = 0$ and $\mathcal{A}(1, \ldots, 1) = 1$

Any aggregation operator $\mathcal{A}$ can be interpreted as a family of mappings $\{A_n : n \in \mathcal{N}\}$ such that:

$$A_n : [0,1]^n \to [0,1] \quad \text{and} \quad A_n(x_1, \ldots, x_n) = \mathcal{A}(x_1, \ldots, x_n)$$

Of course, $A_1 \equiv id_{[0,1]}$.

There are many mappings satisfying the above definition, e. g., the arithmetic mean, the geometrical mean, the harmonic mean, the p-mean, etc. We focus attention on triangular norms being reviewed as aggregating operators.

12.2.3 Finite automata

In this discussion, we consider two types of finite automata: deterministic and nondeterministic ones, as studied in Chapter 7. In this Chapter, we call them *crisp* automata in contrast to automata used to process imperfect information. In the next two sections, we briefly refer to deterministic and nondeterministic crisp automata.

12.2.3.1 Deterministic finite automata

Deterministic finite automata are systems $A = (Q, \Sigma, \delta, q_0, F)$, where: $Q = \{q_0, q_1, \ldots, q_r\}$ – the set of states, $\Sigma = \{a^{(1)}, a^{(2)}, \ldots, a^{(p)}\}$ – the input alphabet, $\delta : Q \times \Sigma \to Q$ – the transition function, $q_0 \in Q$ – the initial state, $F \subset Q$ – the set of final states.

Finite automata accept languages, which are subsets of the set Σ^* of all finite strings over the input alphabet Σ. For an input string $w = a_1 a_2 \cdots a_n$, we define computation realized by the automaton A as the sequence:

$$q_0 \; a_a \; q_1 \; a_2 \; q_2 \; \cdots \; a_{n-1} \; q_{n-1} \; a_n \; q_n \tag{12.2}$$

where successive states of computation are determined by applying the transition function: $q_i = \delta(q_{i-1}, a_i)$ for $i = 1, 2, \ldots, n$. We say that the string w is accepted by the automaton A if and only if $q_n \in F$.

12.2.3.2 Nondeterministic finite automata

The definition of nondeterministic automata differs in the transition function when compared with the above definition of deterministic automata: $\delta : Q \times \Sigma \to 2^Q$ is the transition function of nondeterministic automata, where 2^Q is the family of all subsets of the set of states Q. Note that both classes of finite automata, i. e., deterministic and nondeterministic, are equivalent, that is, for any automaton from any class there exists some automaton of the other class accepting the same language, cf. [5].

Let us have a look at computation of nondeterministic finite automata for the input string $w = a_1 a_2 \cdots a_n$. In comparison with the deterministic case, instead of single states as in (12.2), we obtain sets of states shown below:

$$Q_0 \; a_a \; Q_1 \; a_2 \; Q_2 \; \cdots \; a_{n-1} \; Q_{n-1} \; a_n \; Q_n \tag{12.3}$$

where: $Q_0 = \{q_0\}$ and $Q_i = \bigcup_{p \in Q_{i-1}} \delta(p, a_i)$ for $i = 1, 2, \ldots, n$. We say that the nondeterministic automaton $A = (Q, \Sigma, \delta, q_0, F)$ accepts its input w if and only if $Q_n \cap F \neq \emptyset$.

12.2.4 Finite automata with imperfect information

We consider different types of imperfectness of information in the definition of finite automata with imperfect information, that is, we assume that both states and input symbols could be imperfectly described. On the other hand, we assume that the proposed definition is a generalization of the existing definitions of such automata mentioned above. Therefore, some suitable restrictions have to be considered.

Finite automata with imperfect information are systems:

$$A = (Q, \Sigma, \delta, \lambda_0, \lambda_A) \tag{12.4}$$

where:

$Q = \{q_0, q_1, \ldots, q_r\}$	the set of states,
$\Sigma = \{a^{(1)}, a^{(2)}, \ldots, a^{(\rho)}\}$	the input alphabet,
$\delta : Q \times \Sigma \to \mathbb{Y}(Q)$	the transition function,
$\lambda_0 : Q \to \mathbb{Y}$	the initial states designator,
$\lambda_A : Q \to \mathbb{Y}$	the final states designator,

furthermore:

- $\mathbb{Y}$ identifies a type of information imperfectness, that is, the set $\{0,1\}$ is associated with finite deterministic and nondeterministic automata, the interval $[0,1]$ corresponds to fuzzy automata, etc. We call $\mathbb{Y}$ *imperfectness domain*,
- $\lambda : Q \to \mathbb{Y}$ is a mapping defining states assignment for a given configuration of an automaton, λ is called a states designation,
- $\sigma \equiv \bigcup_{k=0}^{\infty}\{\sigma^k\}$, $\sigma^k : \Sigma^k \to \mathbb{Y}^k$ and $\sigma^0 : \varepsilon \to \mathbb{Y}$ are mappings defining assignments of input strings of length k, σ is called input designation,
- $\mathbb{Y}(Q)$ and $\mathbb{Y}(\Sigma^*)$ denote families of states' and inputs' designators.

Notice that, unlike in automata presented above, we admit input to be a form of imperfect information. For instance, we admit consecutive elements of input sequence in the form of a mapping of the input alphabet Σ into the domain $\mathbb{Y}$.

By analogy to finite automata and fuzzy finite automata, we ensue the computation as the sequence of state designators and input designators. Namely, for the given states designator λ_{i-1} and the given input designator σ_i, the automaton makes a move switching to the states designator λ_i. So then, for the given input $\omega = \sigma_1\sigma_2\ldots\sigma_n$, computation of an automaton is described by the following sequence of states and input designators:

$$\lambda_0\ \sigma_1\ \lambda_1\ \sigma_2\ \lambda_2\ \sigma_3\ \ldots\ \sigma_{n-1}\ \lambda_{n-1}\ \sigma_n\ \lambda_n \tag{12.5}$$

For given transition function δ we employ the step function Δ to compute consecutive states designators λ_i the automaton computation for the given input as above:

$$\begin{aligned} &\Delta : \mathbb{F}(Q) \times \mathbb{F}(\Sigma) \to \mathbb{F}(Q) \\ &\lambda_i = \Delta(\lambda_{i-1}, \sigma_i) \quad \text{for } i = 1, 2, \ldots, n \end{aligned} \tag{12.6}$$

The detailed formula for computing successive states designators produces a degree, to which every state belongs to given states designators, i. e., for given λ_i and for any $p \in Q$:

$$\lambda_i(p) = \Delta(\lambda_{i-1}, \sigma_i)(p) = \underset{q\in Q, a\in\Sigma}{\mathrm{S}}(\mathrm{T}(\lambda_{i-1}(q), \sigma_i(a), \delta(q,a)(p))) \tag{12.7}$$

Acceptance of a finite automaton with imperfect information is the degree to which the output states designator is compatible with the final states designators. Namely, the given input ω is accepted to the degree:

$$\mu(\omega) = \underset{q\in Q}{\mathrm{S}}(\mathrm{T}(\lambda_n(q), \lambda_A(q))) \tag{12.8}$$

12.2.5 Time series and their representation

Representation of a time series, in terms of an FCM-based model architecture (i. e., a fuzzy cognitive maps-based model FCM), can be decomposed into two essential design stages: a way to capture the dynamics of the system and extract concepts – nodes in the map.

Concepts can be seen as aggregates of information, whose specificity determines the accuracy of the description of phenomena. In this light, FCMs align with the idea of granular computing, which is focused on knowledge granules – abstract units of information. Nodes are knowledge granules, which in FCMs are typically represented by fuzzy sets.

The proposed approach is illustrated here using several synthetic time series. We deliberately start with synthetic time series because of their illustrative properties. The considered time series are built involving two sets of numbers of cardinality 3 and 5: $\{2, 5, 8\}$ and $\{1, 3, 5, 7, 9\}$ and one time series is built on a set of three numbers $\{2, 6, 8\}$. A procedure for synthesizing a time series is:

1. select base sequence, which is aimed to cover different numbers of concepts in the three dimensions: values, value changes and changes of value changes. This aim is gained by a different order of base sequence elements, refer to Table 12.2,
2. replicate base sequence so that the total length of a time series is 3000,
3. add random noise drawn from the normal (Gaussian) probability distribution with the mean equal to 0 and standard deviation equal to 0.7.

With this procedure, we can obtain time series of varying complexity, depending on the complexity of the base sequence assumed in the first step. In this way, we capture the quantitative performance of a given FCM for different configurations of input time series. Owing to the Gaussian noise, such a time series exerts both regular, periodic variations and random variations.

Figure 12.1 visualizes the idea behind the proposed approach. Both plots concern one of the easiest time series we addressed in this study. It was constructed based on a sequence of numbers 2, 6, 8. As a result, in this particular time series, we can clearly distinguish three one-dimensional concepts that could ideally represent its amplitude. Visual inspection of Figure 12.1 confirms that even without any knowledge about the

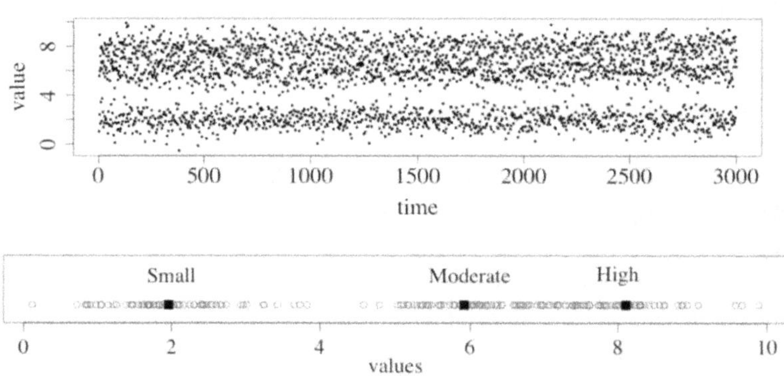

Figure 12.1: The synthetic time series based on period 268 in time/amplitude coordinates (upper part) and first 200 elements on the amplitude axis (bottom part).

specifics of the 3-step procedure listed above, the dataset is clearly composed of the three dense clusters of data.

To enhance the human-centric nature of an FCM model, we usually assign linguistic labels. In the case above, we can use "Small" to represent values centered around 2, "Moderately High" to represent clusters related to 6, and "High" to define values close to 8. For convenience, we abbreviated the linguistic variables to S, MH, and H, respectively.

A dot-plot of amplitude in time (3000 time points) is shown in the upper part of Figure 12.1, while the bottom plot shows the first 200 points on the amplitude axis with centers of amplitude groups.

It is apparent that in real-world data, it is much harder to detect such regularities. Hence, among the biggest challenges in the domain of FCM-based modelling is to propose, so to speak, a bulletproof method for concept extraction that will work even when the data is highly irregular. Please note that concept extraction, although it may resemble clustering at first sight, is a domain-specific task. Concepts in an FCM generalize knowledge. A fundamental parameter of an FCM related to the number of its concepts is its specificity. The larger the map, the higher its specificity. We expect specific maps to be numerically more accurate than less particular maps. At the same time, very specific maps are much more difficult to interpret than less specific ones.

Let us continue with the main course: time series representation.

12.2.5.1 Dynamics based representation

Time series is a sequence of numbers, say *amplitudes*. Let us assume that we have a time series of length $N + 2$

$$a_{-1}, a_0, a_1, a_2, a_3, \dots, a_N$$

To capture the dynamics of the system, we propose switching to coordinates *amplitudes/amplitude changes/changes of amplitude change* for consecutive time points forming the following series of triples:

$$(a_1, da_1, dda_1), (a_2, da_2, dda_2), \dots, (a_N, da_N, dda_N)$$

where for any triple (a_i, da_i, dda_i) the change of amplitude is $da_i = a_i - a_{i-1}$ and change of amplitude change is $dda_i = da_i - da_{i-1} = a_i - 2 \cdot a_{i-1} + a_{i-2}$ for $i = 1, 2, \dots, N$.[1] Note that the original time series are padded with the preceding elements a_{-1}, a_0 to allow computing amplitude change and change of amplitude change for all time points $i = 1, 2, \dots, N$.

An example of such a series representation is in Table 12.2. It concerns the already mentioned 2, 6, 8-based synthetic time series.

1 Note, this representation corresponds to *function/first derivative/second derivative* in differential calculus.

Table 12.2: An example of time series dynamics-based representation built on the sequence 268 (non-distorted values).

time	−1	0	1	2	3	4	5	6	7	8	...	2998	2999	3000
amplitudes	6	8	2	6	8	2	6	8	2	6	...	2	6	8
amplitude changes	~	2	−6	4	2	−6	4	2	−6	4	...	−6	4	2
changes of amplitude change	~	~	−8	10	−2	−8	10	−2	−8	10	...	−8	10	−2

In Table 12.2, we display the values affected by some random noise. The first row refers to consecutive time points, i. e., enumerated numbers. The second row contains numerical amplitude values. Notice that amplitudes are padded with the first values 6 and 8 (corresponding to time −1 and 0) to allow computation of amplitude change and change of amplitude change in time 1. The third and fourth rows correspond to amplitude changes and changes in amplitude change. Therefore, we have the following triples in the dynamics-based representation:

$$(2,-6,-8),(6,4,10),(8,2,-2),(2,-6,-8),(6,4,10),(8,2,-2),\dots$$

12.2.5.2 History based representation

Dynamics-based representation of time series obviously is conceptually equivalent to the series of triples of *present/past/before past* amplitudes:

$$(a_1,a_0,a_{-1}),(a_2,a_1,a_0),\dots,(a_N,a_{N-1},a_{N-2})$$

This representation of the series, based on the sequence 2, 6, and 8, is shown in Table 12.3. By analogy, we have the following triples in the history-based representation:

$$(2,8,6),(6,2,8),(8,6,2),(2,8,6),(6,2,8),(8,6,2),\dots$$

Theoretically, both representations are equivalent as long as they represent a time span of the same length, i. e.:

- dynamics-based representation with (amplitudes, amplitude changes) pairs is equivalent to (present, past) pairs in history-based representation,

Table 12.3: An example of a time series history-based representation built on the sequence 268 (non-distorted values).

time	−1	0	1	2	3	4	5	6	7	8	9	...	2998	2999	3000
present amplitudes	6	8	2	6	8	2	6	8	2	6	8	...	2	6	8
past amplitudes	~	~	8	2	6	8	2	6	8	2	6	...	8	2	6
before past amplitudes	~	~	6	8	2	6	8	2	6	8	2	...	6	8	2

- dynamics-based representation with (amplitudes, amplitude changes, changes of amplitude change) triples is equivalent to (present, past, before past)triples in history-based representation,
- and so on.

In what follows, we elaborate on the time series representation, which captures three data points.

Note that the longer the time span we capture, the more complex the model gets. If the objective of modelling is to build a model to be presented and interpreted by a human being, a time span of 3 data points is among the highest reasonable values. Representation based on triples corresponds to a 3-dimensional space. It means that with such representation, we can plot the time series in a 3-dimensional coordinate system of either (amplitudes, amplitude changes, changes in amplitude change) coordinates or in (present, past, before past) coordinates. 3-dimensional space is the highest system a human being can easily visualize.

In the sequel, we elaborate on the representation of time series and concepts in such 3-dimensional spaces.

A commonly encountered research demonstrates that the higher dimensionality we use for time series representation, the lower the numerical error of prediction becomes. At the same time, increasing dimensionality entails significant drawbacks, including computational and interpretational overhead. The 3-dimensional space can be thought of as a reasonable compromise between numerical accuracy and interpretability.

12.3 Illustrative examples

Computation realized by different types of automata for potentially infinite input series of signals is discussed in this section. The input alphabet Σ is restricted to two elements, and the series of input signals is strictly structured. Such a series may have admitted irregularities and not admitted abnormalities. Automata are designed to detect and report both irregularities and abnormalities.

Let us consider time series over the alphabet $\{a, b\}$. The regular series of signals are double *ones* separated by singe *zeros*, e. g. the following series is correct:

$$w_1 = \dots bbabba \dots$$

There might be irregularities of two *zeros* and three *ones* in such sequences, e. g. the following series are also admitted:

$$w_2 = \dots abbaabbaabbbabb \dots$$
$$w_3 = \dots bbbaabbabbbaabb \dots$$

On the other hand, the following series exhibits abnormality:

$$w_4 = \dots bbabbbaaabb\dots$$

In the following sections, methods for reporting irregularities and abnormalities by different types of automata are illustrated.

12.3.1 Classical finite automata

12.3.1.1 Deterministic case

We refer to the description of deterministic finite automata given in Section 12.2.3.1. An automaton for processing the above series of signals w_1, w_2, w_3, w_4 is constructed. Transition diagram of a simple finite automaton accepting such a series of signals is presented in Figure 12.2. The design of this automaton was arbitrarily invented. For the sake of simplicity, we do not include the initial state since it is used only once to initiate computation and is never visited. The automaton switches from the initial state to the state either q_a or q_b for input signal either a or b, respectively. Namely, denoting the initial state by $\rightarrow q$ we get the following values of the transition function: $\delta(\rightarrow q, a) = q_a$ and $\delta(\rightarrow q, b) = q_b$. Therefore, the formal description of the automaton is $A = (Q, \Sigma, \delta, \rightarrow q, F)$, where $Q = \{\rightarrow q, q_a, q_{aa}, q_b, q_{bb}, q_{bbb}, q_R\}$, $F = \{q_a, q_{aa}, q_b, q_{bb}, q_{bbb}\}$. The transition table of this automaton is given in Table 12.4. In this table, the initial state $\rightarrow q$ is dropped, so then we use the set of states $Q = \{q_a, q_{aa}, q_b, q_{bb}, q_{bbb}, q_R\}$.

This simple automaton falls into rejecting state q_R once an irregularity appears, then it stays in this state infinitely.

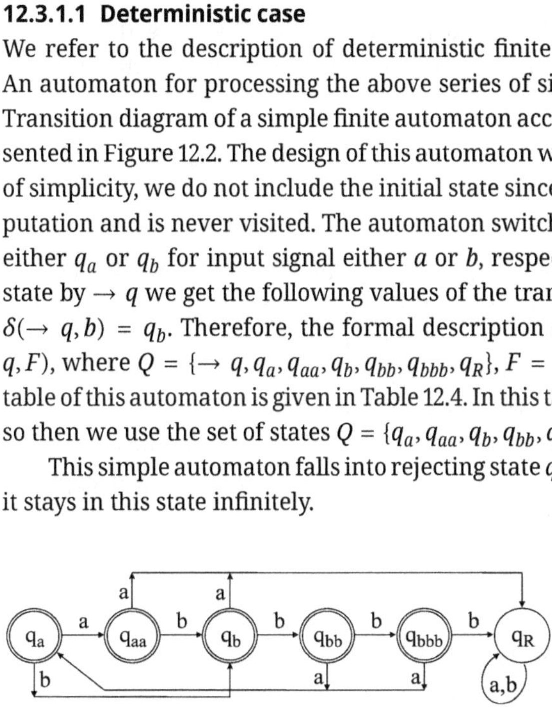

Figure 12.2: Transition diagram of the finite deterministic automaton. Accepting states are double circled.

Table 12.4: Transition table of the finite deterministic automaton. Right arrows going out of states denote accepting states.

δ	a	b
$q_a \rightarrow$	q_{aa}	q_b
$q_{aa} \rightarrow$	q_R	q_b
$q_b \rightarrow$	q_R	q_{bb}
$q_{bb} \rightarrow$	q_a	q_{bbb}
$q_{bbb} \rightarrow$	q_a	q_R
q_R	q_R	q_R

Computations for input series of signals defined above: $w_1 = \dots bbabba \dots$ and $w_2 = \dots abbaabbabbbabb \dots$ are as follows:

w_1: $\dots b\, q_b\, b\, q_{bb}\, a\, q_a\, b\, q_b\, b\, q_{bb}\, a\, q_a \dots$

w_2: $\dots a\, q_a\, b\, q_b\, b\, q_{bb}\, a\, q_a\, a\, q_{aa}\, b\, q_b\, b\, q_{bb}\, a\, q_a\, a\, q_{aa}\, b\, q_b\, b\, q_{bb}\, b\, q_{bbb}\, a\, q_a\, b\, q_b\, b\, q_{bb} \dots$

In both cases, computation visits accepting states since no abnormality has been encountered. Irregularities are treated like a regular series of signals. Of course, we can build a deterministic automaton reporting irregularities, but it raises a significant increase in the set of states.

The series $w_4 = \dots bbabbbaaabb \dots$ launches abnormality. Therefore, the automaton, once it meets abnormality, falls into the rejecting "trap" state q_R and then stays in this state infinitely:

$$w_4: \quad \dots b\; q_b\; b\; q_{bb}\; a\; q_a\; b\; q_b\; b\; q_{bb}\; b\; q_{bbb}\; a\; q_a\; a\; q_{aa}\; a\; q_R\; b\; q_R\; b\; q_R \dots$$

12.3.1.2 Nondeterministic case

Referring to the description of nondeterministic finite automata given in Section 12.2.3.2, the transition diagram of a simple nondeterministic finite automaton is presented. The diagram is displayed in Figure 12.3. This automaton is equivalent to the above deterministic one, i. e., it accepts the same series of input signals. The formal description of the automaton is the same as in the case of deterministic one, i. e. $A = (Q, \Sigma, \delta, \rightarrow q, F)$, where $Q = \{\rightarrow q, q_a, q_{aa}, q_b, q_{bb}, q_{bbb}, q_R\}$, $F = \{q_a, q_{aa}, q_b, q_{bb}, q_{bbb}\}$. The transition function of this automaton is given in Table 12.5. Again, the initial state $\rightarrow q$ is dropped in the transition table, so then the set of states $Q = \{q_a, q_{aa}, q_b, q_{bb}, q_{bbb}, q_R\}$ is used there.

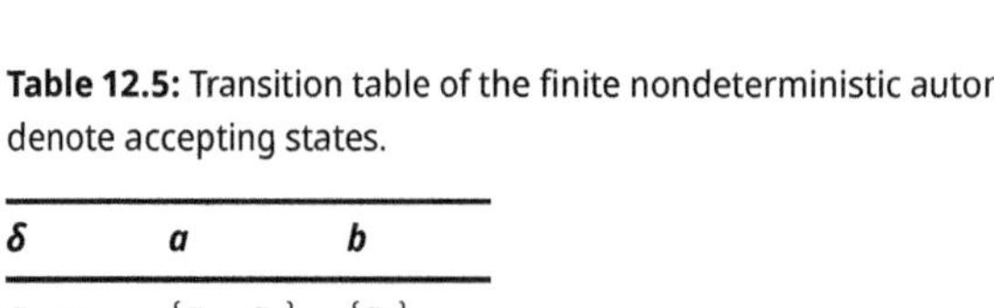

Figure 12.3: Transition diagram of the finite deterministic automaton. Accepting states are double circled.

Table 12.5: Transition table of the finite nondeterministic automaton. Right arrows going out of states denote accepting states.

δ	a	b
$q_a \rightarrow$	$\{q_{aa}, q_R\}$	$\{q_b\}$
$q_{aa} \rightarrow$	$\{q_R\}$	$\{q_b\}$
$q_b \rightarrow$	$\{q_R\}$	$\{q_{bb}\}$
$q_{bb} \rightarrow$	$\{q_a\}$	$\{q_{bbb}, q_R\}$
$q_{bbb} \rightarrow$	$\{q_a\}$	$\{q_R\}$
q_R	$\{q_R\}$	$\{q_R\}$

Now, let us consider computations of the nondeterministic automaton for input series defined above: $w_1 = \ldots bbabba \ldots$ and $w_2 = \ldots abbaabbabbbabb \ldots$. These computations are carried out as follows:

w_1: $\ldots b\ \{q_b\}\ b\ \{q_{bb}\}\ a\ \{q_a\}\ b\ \{q_b\}\ b\ \{q_{bb}\}\ a\ \{q_a\} \ldots$

w_2: $\ldots a\ \{q_a\}\ b\ \{q_b\}\ b\ \{q_{bb}\}\ a\ \{q_a\}\ a\ \{q_{aa}, q_R\}\ b\ \{q_b, q_R\}\ b\ \{q_{bb}, q_R\}\ a\ \{q_a, q_R\}\ a\ \{q_{aa}, q_R\}$
$b\ \{q_b, q_R\}\ b\ \{q_{bb}, q_R\}\ b\ \{q_{bbb}, q_R\}\ a\ \{q_a, q_R\}\ b\ \{q_b, q_R\}\ b\ \{q_{bb}, q_R\} \ldots$

In both cases, computation visits accepting states since no abnormality has been encountered. However, unlike in the deterministic case, any irregularity is detected and signalized by the rejecting "trap" state being an elements of computation. Note that once an irregularity is encountered, computation stays in rejecting states infinitely. No further irregularity affects the computation of this automaton since the irregularity report is idempotent. The computation remains in accepting states until an abnormality is encountered.

As in the deterministic case, the series of input signals $w_4 = \ldots bbabbbaaabb \ldots$ launches abnormality, therefore the automaton falls into the rejecting "trap" state q_R once it meets abnormality and then stays in this state infinitely:

w_4: $\ldots b\ \{q_b\}\ b\ \{q_{bb}\}\ a\ \{q_a\}\ b\ \{q_b\}\ b\ \{q_{bb}\}\ b\ \{q_{bbb}, q_R\}\ a\ \{q_a, q_R\}\ a\ \{q_{aa}, q_R\}\ a\ \{q_R\}$
$b\ \{q_R\}\ b\ \{q_R\} \ldots$

12.3.2 Generalized description of finite automata

12.3.2.1 Deterministic case

Here, let us rewrite the transition table given in Table 12.4 in the spirit of formula (12.4). This rewriting illustrates how automata with imperfect information operate. In rows of Table 12.6 we have mappings from the set of states $Q = \{q_a, q_{aa}, q_b, q_{bb}, q_{bbb}, q_R\}$ into states designators valued in the crisp unipolar scale $\{0, 1\}$. For instance, the state q_a in Table 12.4 is replaced by the state designator $(0, 1, 0, 0, 0, 0)$ (of the mapping $\lambda : Q \to \{0, 1\}$) in Table 12.6. In fact, formally, this automaton and the automaton defined in Section 12.3.2.2 are fuzzy finite automata, cf. [3].

Computation of the automaton can be restructured in the spirit of formulas (12.5) and (12.7) in a similar way. The design of this automaton is based on operators max and min, which are employed in place of S and T operators of formulas (12.7) and (12.8), respectively. Input designators σ are shown in the left column of this table, and state designators λ are given in the right column of the table. Finally, in virtue of formula (12.8), the final states designator is defined as $(1, 1, 1, 1, 1, 0)$. So then, according to formula (12.8), the automaton accepts a string as long as computation visits final states. Computation for the input w_1 is given in Table 12.7.

Table 12.6: Transition table of the deterministic automaton from Figure 12.2 and Table 12.4.

δ	q_a	q_{aa}	q_b	q_{bb}	q_{bbb}	q_R	q_a	q_{aa}	q_b	q_{bb}	q_{bbb}	q_R
			a						b			
q_a	0	1	0	0	0	0	0	0	1	0	0	0
q_{aa}	0	0	0	0	0	1	0	0	1	0	0	0
q_b	0	0	0	0	0	1	0	0	0	1	0	0
q_{bb}	1	0	0	0	0	0	0	0	0	0	1	0
q_{bbb}	1	0	0	0	0	0	0	0	0	0	0	1
q_R	0	0	0	0	0	1	0	0	0	0	0	1

Table 12.7: Computation of the automaton with transition table given in Table 12.6 for the input $w_1 = \ldots bbabba \ldots$.

a_i	b_i	q_a	q_{aa}	q_b	q_{bb}	q_{bbb}	q_R
σ_i				λ_i			
...		...					
0	1	0	0	1	0	0	0
0	1	0	0	0	1	0	0
1	0	1	0	0	0	0	0
0	1	0	0	1	0	0	0
0	1	0	0	0	1	0	0
1	0	1	0	0	0	0	0
...		...					

12.3.2.2 Nondeterministic case

Here we consider a general description of the nondeterministic automaton defined in Section 12.3.1.2. Transition table of this automaton is given in Table 12.8. As above, the design of this automaton is based on operators max and min, which are employed in place of operators S and T of formulas (12.7) and (12.8), respectively.

This automaton "traps" irregularities: the second a and the third b at input. Notice that irregularities switch computation to the rejecting "trap" state q_R, but until the input series does not contain abnormalities, computation is not taken out of the accepting states, i. e., it stays in both accepting states and the rejecting "trap" one. Therefore, we can test input structures using the final (accepting) states designator $(1, 1, 1, 1, 1, 0)$ and the rejecting states designator $(0, 0, 0, 0, 0, 1)$, applying formula (12.8) in both cases.

The computation for the series w_4 is given in Table 12.9. Notice that the 6th input symbol is irregular. For this case, the following two symbol computations remain in two states: an accepting one and a rejecting "trap" one. Then, the 9th symbol is abnormal, which switches computation leading to the rejecting state only. Computation stays in the rejecting state infinitely.

Table 12.8: Transition table of the nondeterministic automaton from Figure 12.3 and Table 12.5.

	q_a	q_{aa}	q_b	q_{bb}	q_{bbb}	q_R	q_a	q_{aa}	q_b	q_{bb}	q_{bbb}	q_R
δ			a						b			
q_a	0	1	0	0	0	1	0	0	1	0	0	0
q_{aa}	0	0	0	0	0	1	0	0	1	0	0	0
q_b	0	0	0	0	0	1	0	0	0	1	0	0
q_{bb}	1	0	0	0	0	0	0	0	0	0	1	1
q_{bbb}	1	0	0	0	0	1	0	0	0	0	0	1
q_R	0	0	0	0	0	1	0	0	0	0	0	1

Table 12.9: Computation of the nondeterministic automaton with transition table given in Table 12.8 for the input $w_4 = \ldots bbabbbaaabb \ldots$.

a_i	b_i	q_a	q_{aa}	q_b	q_{bb}	q_{bbb}	q_R
σ_i				λ_i			
...		...					
		1	0	0	0	0	0
0	1	0	0	1	0	0	0
0	1	0	0	0	1	0	0
1	0	1	0	0	0	0	0
0	1	0	0	1	0	0	0
0	1	0	0	0	1	0	0
0	1	0	0	0	0	1	1
1	0	1	0	0	0	0	1
1	0	0	1	0	0	0	1
1	0	0	0	0	0	0	1
0	1	0	0	0	0	0	1
0	1	0	0	0	0	0	1
...		...					

12.3.3 Cumulating irregularities

Now let us consider cumulation irregularities: the second a and the third b at input. We apply an automaton with imperfect information for the detection of such irregularities. In fact, imperfectness appears in the transition function, where we have the second a and third b transition to the state q_R with the membership 0.25. The transition table of such an automaton is given in Table 12.10. Operators S and T standing in formulas (12.7) and (12.8) are replaced by bounded sum t-conorm and Lukasiewicz t-norm, cf. Table 12.1. Note, this choice was completely arbitrary.

As in the case of a nondeterministic automaton, the final states designator is equal to $(1, 1, 1, 1, 1, 0)$. We can also use the rejecting states designator equal to $(0, 0, 0, 0, 0, 1)$ to get parallel reports on regularity/irregularity/abnormality of input data. Notice that,

Table 12.10: Transition table of the automaton with cumulation of irregularities.

δ	q_a	q_{aa}	q_b	q_{bb}	q_{bbb}	q_R	q_a	q_{aa}	q_b	q_{bb}	q_{bbb}	q_R
				a						*b*		
q_a	0	1	0	0	0	.25	0	0	1	0	0	0
q_{aa}	0	0	0	0	0	1	0	0	1	0	0	0
q_b	0	0	0	0	0	1	0	0	0	1	0	0
q_{bb}	1	0	0	0	0	0	0	0	0	0	1	.25
q_{bbb}	1	0	0	0	0	0	0	0	0	0	0	1
q_R	0	0	0	0	0	1	0	0	0	0	0	1

Table 12.11: Computation of the automaton with transition table given in Table 12.10 for the input w_3 = ...*bbabbbaaabb*....

a_i	b_i	q_a	q_{aa}	q_b	q_{bb}	q_{bbb}	q_R
σ_i				λ_i			
...		...					
		1	0	0	0	0	0
0	1	0	0	1	0	0	0
0	1	0	0	0	1	0	0
0	1	0	0	0	0	1	0.25
1	0	1	0	0	0	0	0.25
1	0	1	0	0	0	0	0.50
0	1	0	0	1	0	0	0.50
0	1	0	0	0	1	0	0.50
1	0	1	0	0	0	0	0.50
0	1	0	0	0	0	1	0.50
0	1	0	0	0	0	1	0.50
0	1	0	0	0	0	1	0.75
1	0	0	1	0	0	0	0.75
1	0	0	0	0	0	0	1
0	1	0	0	0	0	0	1
0	1	0	0	0	0	0	1
...		...					

unlike in the nondeterministic case, we can observe the accumulation of irregularities, which ultimately leads to reporting the state of abnormality, cf. Table 12.11.

Finally, in Table 12.12 a computation example with imprecise input is presented. The fourth and seventh input data are affected by noise. For the automaton being discussed, this is a kind of irregularity. This irregularity is accumulated in the rejection state q_R.

Table 12.12: Computation of the automaton with transition table given in Table 12.10 for an imperfect input.

a_i	b_i	q_a	q_{aa}	q_b	q_{bb}	q_{bbb}	q_R
σ_i		λ_i					
...		...					
		1	0	0	0	0	0
0	1	0	0	1	0	0	0
0	1	0	0	0	1	0	0
0.25	1	0	0	0	0	1	0.25
1	0	1	0	0	0	0	0.25
1	0	1	0	0	0	0	0.75
0	1	0	0	1	0	0	0.75
0.25	1	0	0	0	1	0	1
1	0	1	0	0	0	0	1
0	1	0	0	0	0	1	1
0	1	0	0	0	0	1	1
0	1	0	0	0	0	1	1
1	0	0	1	0	0	0	1
1	0	0	0	0	0	0	1
0	1	0	0	0	0	0	1
0	1	0	0	0	0	0	1
...		...					

12.4 Conclusions

The proposed procedure for space similarity evaluation takes advantage of the underlying partition of the given two spaces into disjoint subsets. Such structuring is commonly observed in nature – phenomena from the same universe are divided into groups. We take advantage of such division, assuming that the structuring of spaces is a subject of similarity. The presented formulas for similarity calculation are inspired by human cognitive processes and express rather pessimistic evaluations by considering selected aggregation operators.

The objective was to introduce some underlying methodology. We aimed to focus on proposed similarity relations. A number of illustrative examples were provided to highlight the key ideas of computing.

In future research, this direction could be continued. Similarities of structured spaces with alternative definitions of the structuring would be a choice, for example, inclusions.

Bibliography

[1] Belohlavek R., Determinism and fuzzy automata, *Inf. Sci.* 143(1-4), 205–209 (2002).
[2] Gupta S., Ray A., Keller E., Symbolic time series analysis of ultrasonic data for early detection of fatigue damage, *Mech. Syst. Signal Process.* 21, 866–884 (2007).
[3] Homenda W., Pedrycz W., in: N.-T. Nguyen et al. (Eds.), *Finite Automata with Imperfect Information as Classification Tools*, Lecture Notes in Artificial Intelligence, LNAI, vol. 7653, 2012.
[4] Hopcroft J. E., Ullman J. D., *Introduction to Automata Theory, Languages and Computation*, Addison-Wesley Publishing Company, Reading, Massachusetts, 1979, 2001.
[5] Hopcroft J. E., Motwani R., Ullman J. D., *Introduction to Automata Theory, Languages and Computation*, Addison-Wesley, Boston, 2001.
[6] Kandel A., Lee S. C., *Fuzzy Switching and Automata: Theory and Applications*, Crane Russak & Co, 1979, 2007.
[7] Klement E. P., Mesiar R., Pap E., *Triangular Norms*, Kluwer Academic Publishers, Dordrecht, 2000.
[8] Mordeson J. N., Malik D. S., *Fuzzy Automata and Languages: Theory and Applications*, Taylor & Francis, 2010.

Additional readings

[9] Aho A. V. , Hopcroft J. E., Ullman J. D., *Data Structures and Algorithms*, Addison-Wesley, Reading, 1983.
[10] Aho A. V., Lam M. S., Sethi R., Ullman J. D., *Compilers: Principles, Techniques, and Tools*, 2nd edition, Pearson, 2007.
[11] Aho A. V., Sethi R. S., Ullman J. D., *Compilers: Principles, Techniques and Tools*, Addison-Wesley, Reading, 1986.
[12] Bovet P. B., Crescenzi P., *Introduction to the Theory of Complexity*, Prentice Hall, 2006.
[13] Chomsky N., *Rules and Representations*, Columbia University Press, New York, 1980.
[14] Chomsky N., Halle M., *The Sound Patterns of English*, Harper and Row, New York, 1968.
[15] Chomsky N., *Aspects of a Theory of Syntax*, MIT Press, Cambridge, Massachusetts, 1965.
[16] Chomsky N., *Cartesian Linguistics: A Chapter in the History of Rationalist Thought*, Harper and Row, New York, 1965.
[17] Cormen T., Leiserson C., Rivest R., Stein C., *Introduction to Algorithms*, MIT Press, 2009.
[18] Kinber E., Smith C., *Theory of Computing: A Gentle Introduction*, Sacred Heart University, University of Maryland, Pearson, 2001.
[19] Moret M. B., *The Theory of Computation*, Addison-Wesley Publishing Company, 1998.
[20] Papadimitriou C. H., *Computational Complexity*, Addison-Wesley Longman, 1995.
[21] Sedgewick R., Wayne K., *Algorithms*, Addison-Wesley Professional, 2011.
[22] Shoenfiled J. R., The mathematical work of S. C. Kleene, *Bull. Symb. Log.* (1995).
[23] Sipser M., *Introduction to the Theory of Computation*, Thomson, 2006.
[24] Wilson R. J., *Introduction to Graph Theory*, Addison Wesley, 1996; Pearson Higher Education, 2010.
[25] Wirth N., *Algorithms and Data Structures*, Prentice Hall, 1985.

https://doi.org/10.1515/9783112223123-017

Index

https://doi.org/10.1515/9783112223123-018

www.ingramcontent.com/pod-product-compliance
Lightning Source LLC
Chambersburg PA
CBHW061205240926
51846CB00006B/1115